No Metal No Magic

Element 107

Bohrium, Presented By Blaadlak, From The Magical Elements of the Periodic Table Book Series

Blaadlak

Bohrium

By Sybrina Durant with Illustrations by Pranavva et al.

Bohrium, Presented By Blaadlak
From The Magical Elements of the Periodic
Table Book Series

Story copyright 2026

Soft Cover Print ISBN 13 — 978-1-942740-61-2

BISAC Codes:

JNF051070 JUVENILE NONFICTION / Science & Nature / Chemistry

JNF016000 JUVENILE NONFICTION / Curiosities & Wonders

JNF051080 JUVENILE NONFICTION / Science & Nature / Earth Sciences / General

Soft Cover Print ISBN 13 — 978-1-942740-61-2

Blaadlak Presents Bohrium

This Element 107 book features the periodic table element, Bohrium. It is presented by Blaadlak, a Radioactive Goblin who is fueled by the powers of this periodic table element.

Blaadlak is just one of the 118 elementals who will present all of the Magical Elements of the Periodic Table to readers who are curious about the wonders of the world.

Blaadlak introduces the very magical element, Bohrium, in his book. And it is the very opposite of "boring".

The Radioactive Goblins and their other techno-magical friends are the perfect group to introduce you to the elements in the Periodic Table. Hopefully, this Magical Elements of the periodic table book will spark an interest in the magical and real world properties of all the elements known today. You may be surprised at how prominently they feature in our every day lives.

Each page in this book contains terms that might not be completely familiar to the reader. Refer to the definitions in the back of the book to get a clear understanding of each meaning.

There is also a fun elemental themed Periodic Table at the back of the book. It features 118 elements presented by fanciful characters like unicorns, dragons, wizards, knights and goblins.. They want you to remember that if there's no metal...there's no magic or technology.

Remember, "No metal – No Magic. . .and No Technology".

It's Techo-Magical.

Note: Sybrina Publishing websites are Sybrina.com and MagicalPTElements.com. Follow sybrinapublishing on Instagram, Magical Elements of the Periodic Table on Facebook, @sybrinad on Pinterest, Sybrina_SPT on Twitter; and Sybrina Durant on LinkedIn.

Bohrium is a Super Heavy—Radioactive Element

Bohrium (atomic number 107) was first synthesized in 1981 by a German team led by Peter Armbruster and Gottfried Münzenberg at the Gesellschaft für Schwerionenforschung (GSI) in Darmstadt, Germany. Schwerionenforschung is pronounced shvair-ee-OH-nen-forsch-oong, meaning "heavy ion research" in German.

Bohrium is predicted to be a solid Transition Metal like Rhenium or Technetium in Group 7.

The shortest isotope half-life for Bohrium is 10 milliseconds and the longest is 2.4 minutes. There are reports of an unconfirmed isotope, bohrium-278 (), which may have a half-life of about 11.5 to 19 minutes.

If scientists could make Bohrium last long enough, theoretical evaluations suggest it will be malleable and ductile, allowing it to be shaped, hammered, or drawn into wires. .

As a metal in Group 7, it is expected to be a solid that can conduct heat and electricity under normal conditions.

While not a typical ferromagnetic material like iron, bohrium is classified within research contexts as a heavy metal with potential for use in specialized applications, including magnetic nanoparticle technology.

LEGEND

Alkali Metals
Alkali Earth Metals
Transition Metals
Post-Transition (or Other Metals)
Metalloids
Non-Metals
Halogens
Noble Gases
Rare Earth Lanthanide Metals
Actinide Metals
Super Heavy—Radioactive

Bohrium Element

Atomic Structure

Super Heavy—Radioactive—Superheavy elements are those elements with a large number of protons in their nucleus. Elements with more than 92 protons are unstable; they decay to lighter nuclei with a characteristic half-life. They do not occur in large quantities (if at all) naturally on earth, and only exist briefly under highly controlled circumstances. They include lawrencium, rutherfordium, dubnium, seaborgium, bohrium, hassium, meitnerium, darmstadtium, roentgenium, copernicium, nihonium, flerovium, moscovium, livermorium, tennessine, and oganesson.

Have you ever heard of elements so heavy and so rare that scientists can only make a few atoms of them at a time? One of those elements is called Bohrium, and it is one of the most fascinating and mysterious substances ever created by human hands. Although it is almost never mentioned in everyday conversation, bohrium represents one of the great achievements of modern chemistry and physics because it helps us explore the limits of the periodic table and the structure of matter itself.

Bohrium is a chemical element with the symbol Bh and the atomic number 107. This means that each atom of bohrium has 107 protons in its nucleus, which gives the element its place on the periodic table. The periodic table is more than just a chart of elements; it is a map of the known building blocks of the universe. Bohrium sits near the far end of that map, in a region filled with extremely heavy elements that are usually unstable and short-lived. These elements are often called superheavy elements because their nuclei contain so many protons and neutrons packed into such a tiny space that they struggle to stay together.

Bohrium is classified as a transition metal, placing it in the same broad family as familiar elements such as iron, nickel, copper, and gold. That classification suggests that, in theory, it should have some of the typical metallic qualities of those elements, such as conductivity and a tendency to form ions. However, bohrium is not like the metals people encounter in daily life. Its enormous atomic weight and intense radioactivity make it far more unusual and far less stable than ordinary metals. In fact, scientists have never been able to collect a visible piece of bohrium or study it in the way they might examine a sample of iron or aluminum. Everything known about it comes from observing just a few atoms at a time before they vanish.

Bohrium was first created in 1981 by a team of scientists working at the GSI Helmholtz Centre for Heavy Ion Research in Darmstadt, Germany. The lead scientist on the team was Peter Armbruster, and he worked with Gottfried Münzenberg and other researchers in the group. Their work was part of a larger scientific effort to create and identify new elements beyond those that occur naturally on Earth. Discovering a new element is not as simple as finding a rock or a mineral. It requires advanced equipment, careful measurements, and often a great deal of patience because the atoms are produced in incredibly small numbers and may survive for only moments. Bohrium's discovery was therefore an important milestone in the history of nuclear science.

The element was named in honor of the famous Danish physicist Niels Bohr, whose ideas transformed our understanding of the atom. Bohr made major contributions to atomic theory in the early 1900s, especially through his model of the atom and his work on quantum theory. Naming an element after him was a way of recognizing the importance of his scientific legacy. It is a special distinction to have an element bear your name, and Bohrium is one of several elements that honor a great scientist. The name also reflects how closely modern discoveries are connected to the earlier foundations of physics and chemistry.

Bohrium does not exist naturally on Earth in any meaningful amount. Any bohrium atoms that may have briefly formed in natural processes would decay almost instantly and would be far too rare to detect. Instead, scientists must create it artificially in a laboratory using powerful machines called particle accelerators. These machines are capable of speeding up atomic nuclei to tremendous velocities and then forcing them to collide with target atoms. When the collision conditions are just right, the nuclei may fuse together and form a new, heavier atom. This process is extraordinarily difficult, because the forces involved are immense and the chances of creating the desired atom are very small.

In the case of bohrium, scientists have typically used chromium atoms and a bismuth target in these experiments. The chromium nuclei are accelerated to very high speeds and directed toward the bismuth atoms. If a fusion event occurs and the resulting nucleus is stable long enough to be detected, researchers can identify it by tracking the particles it emits as it decays. Even when the experiment succeeds, the number of bohrium atoms produced is tiny. Scientists may create only a few atoms in an entire experiment, and sometimes only one or two are detected. This is one reason why the element is so difficult to study. There is simply not enough of it to weigh, hold, or observe directly.

Bohrium is also highly radioactive, which means that its atoms are unstable and break apart on their own. Radioactive decay happens when an atomic nucleus is not balanced properly and releases particles or energy in an attempt to become more stable. For bohrium, that process happens quickly. The most stable known isotope, Bohrium-270, has a half-life of about 61 seconds. A half-life is the amount of time it takes for half of a radioactive sample to decay. In practical terms, this means that after a little more than a minute, half of a tiny sample of Bohrium-270 would already be gone, transformed into other elements or isotopes through decay. Because of this short lifespan, bohrium is not something that can be stored for future use. It is an element that seems to appear only briefly, almost like a scientific flash in the dark.

Despite its dramatic properties, bohrium has no practical uses outside of scientific research. It cannot be used in construction, electronics, medicine, or manufacturing because there is too little of it and it disappears too quickly. Its value lies instead in what it teaches scientists. By studying bohrium and other superheavy elements, researchers learn more about nuclear structure, atomic stability, and the forces that hold nuclei together. These studies also help scientists test theories about the periodic table and explore whether there may be an "island of stability," a region where some superheavy elements might last longer than expected. Bohrium is part of this larger scientific journey, helping researchers understand whether even heavier and more stable elements might someday be discovered.

Although bohrium may seem remote from ordinary life, it is a powerful reminder of human curiosity and scientific ingenuity. To create just a few atoms of an element that has never been found naturally on Earth is an extraordinary accomplishment. It shows how far science has advanced and how much remains to be explored. Bohrium may be short-lived, but its discovery has lasting importance because it expands our understanding of the universe at its most fundamental level. In that sense, bohrium is not just one of the heaviest elements on Earth; it is also one of the clearest symbols of the human desire to uncover the unknown.

Bohrium's Family

When you first look at the periodic table of elements it can seem like a giant puzzle filled with symbols, numbers, and strange arrangements. But the periodic table is really a carefully designed map. It helps scientists understand how elements behave, even when those elements are so difficult to study that they barely exist long enough to observe. One of the most challenging of these elements is bohrium, a super-heavy metal that seems to follow the same family patterns as the elements above it.

Bohrium is not found in nature. It is a synthetic element, which means scientists must create it artificially in a laboratory. After bohrium is created, it does not last very long. It is highly radioactive, which means its atoms are unstable and break apart quickly. In fact, bohrium may survive for only a few seconds before disappearing into lighter elements. Because of this, scientists cannot collect a sample in a vial, hold it in their hands, or place it on a scale. They cannot watch it sit in air, melt, or react in a normal way. Instead, they must use the periodic table to predict what it will do.

The periodic table is arranged in rows and columns, and the columns are called groups. Elements in the same group are like members of a family. They may not be identical, but they often share important traits, especially in the way they react with other substances. Just as family members can share similar features, elements in the same group often behave in similar ways because their atoms have similar outer electrons. Bohrium belongs to Group 7, along with manganese, technetium, and rhenium. Of these, rhenium is the element directly above bohrium on the table, so scientists expected bohrium to behave a lot like rhenium, just heavier and more extreme. That was the best guess available before any real experiment could be done.

Of course, testing that guess was not easy. How do you study a metal that vanishes almost immediately after it is made? Scientists had to be extremely quick and very clever. They knew that rhenium forms certain compounds when it reacts with gases such as oxygen and chlorine. So they designed an experiment to see whether bohrium would do something similar. They created just a tiny number of bohrium atoms and sent them into a special chamber filled with oxygen and chlorine gas. The atoms moved through the system so fast that the experiment had to be carefully timed and precisely controlled. Even though only a few bohrium atoms were produced, the scientists were able to detect what happened to them. The result was exciting: bohrium reacted in the predicted way. It combined with oxygen and chlorine and formed a gaseous compound, acting much like rhenium does.

This matters for a big reason. Super-heavy elements are not just bigger versions of ordinary atoms. Their nuclei are so large that the electrons around them can be affected in unusual ways. In the heaviest elements, some electrons move incredibly fast, sometimes approaching the speed of light. When that happens, normal chemical behavior can change. Some elements may not follow the patterns that the periodic table suggests. This is one reason why scientists are so interested in elements like bohrium. They want to know whether the periodic table still works at the farthest edges of known matter or whether new rules begin to appear.

Bohrium showed that, at least in this case, the periodic table still gives a reliable answer. Even though bohrium is extremely heavy, radioactive, and short-lived, it still behaves like the other elements in its family. That is a powerful reminder that science often depends on patterns. The periodic table is not just a chart to memorize in school. It is a predictive tool that lets scientists understand atoms they can barely study at all. The story of bohrium shows that even the rarest and most short-lived elements can reveal deep truths about how the universe works. In the end, bohrium proves that nature often follows its own family rules, no matter how strange or extreme those families may seem.

Potential Future Uses For Bohrium

If scientists are ever able to create long-lasting isotopes of Bohrium, the element is predicted to behave somewhat like Rhenium or Technitium. With that in mind, futurists have come up with some potential future uses:

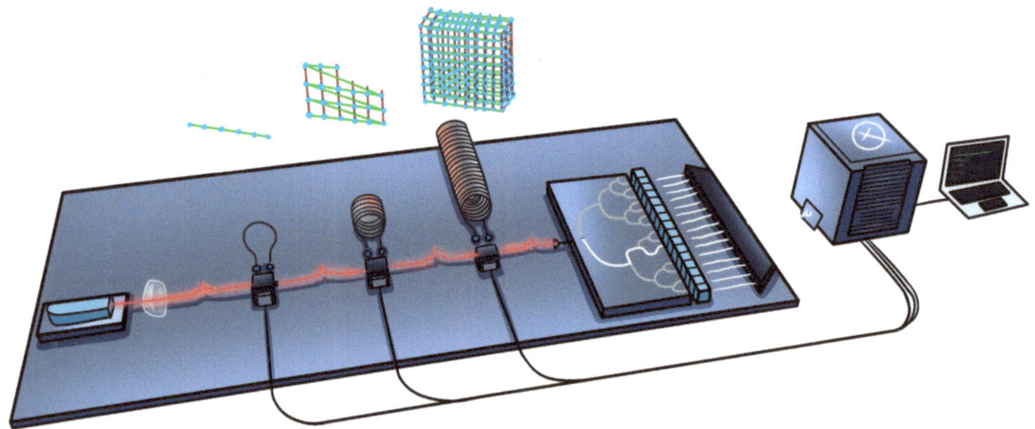

Bohrium isotopes might one day be used in quantum processors, enabling ultra-fast calculations that could transform computing. Their unique nuclear properties may support highly specialized qubits, improve stability, or enhance experimental control in advanced systems. If researchers can safely produce and manipulate these isotopes, they may contribute to breakthroughs in materials science, cryptography, and complex simulations.

A future isotope of Bohrium could potentially serve as a high-energy fuel source for advanced experimental reactors. Because Bohrium is a synthetic, highly unstable element, any practical use would depend on discovering or engineering an isotope with unusual stability, predictable decay behavior, and sufficient energy release. If such an isotope could be produced and controlled safely, it might offer a valuable new option for research into next-generation reactor designs, nuclear physics experiments, and specialized energy systems.

Potential Future Uses For Bohrium

(Continued)

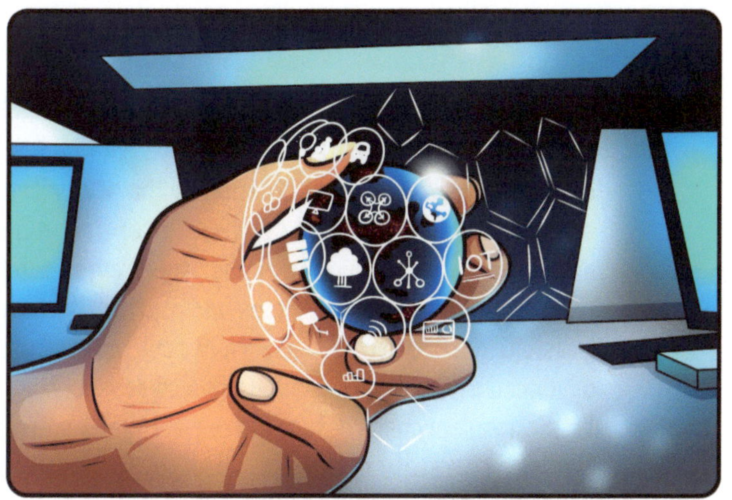

In the future, scientists may develop innovative methods that allow Bohrium nanoparticles to be used in nanomedicine for highly precise cancer treatment. These tiny particles could be engineered to seek out and attach to cancer cells, helping deliver medicine directly to the tumor while reducing harm to healthy tissue. With further research, Bohrium nanoparticles may become an advanced tool for improving the accuracy, effectiveness, and safety of cancer therapies.

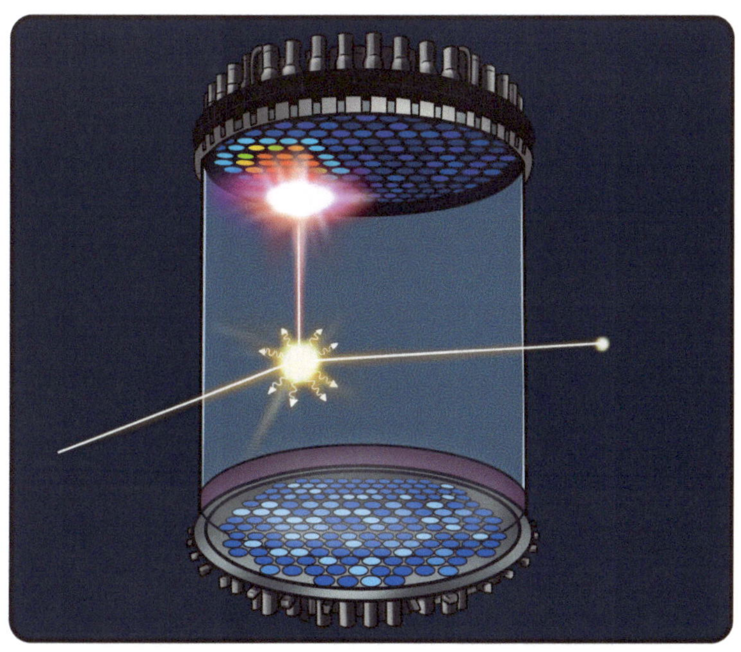

Longer lasting isotopes of bohrium could potentially play an important role in future dark matter research. If such isotopes can be produced and studied, they may help improve the sensitivity and stability of detectors designed to identify extremely rare dark matter interactions. Advances in bohrium isotope research could also contribute to a better understanding of nuclear properties and instrumentation, supporting the development of more precise experimental methods for exploring the universe's hidden matter.

Potential Future Uses For Bohrium

(Continued)

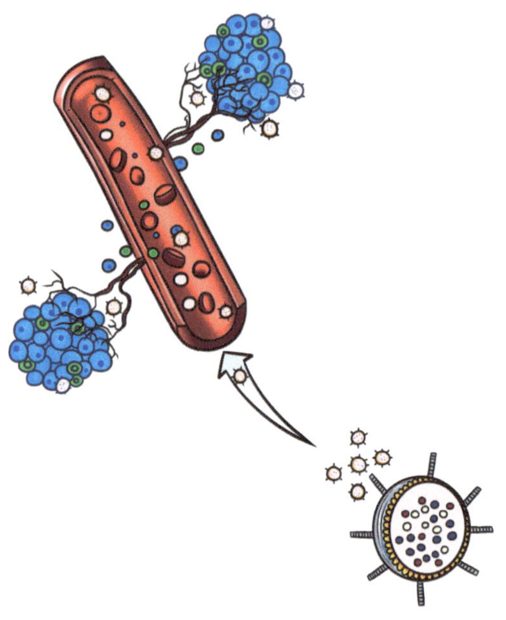

Some day, Bohrium's unusual properties might make it useful in developing highly sensitive particle detectors. Because of its position in the periodic table and its expected chemical behavior, scientists may one day be able to study it in ways that reveal subtle interactions between particles and matter. If these possibilities are realized, Bohrium could contribute to advanced instruments used in physics research, helping researchers detect extremely small signals that are currently difficult to observe.

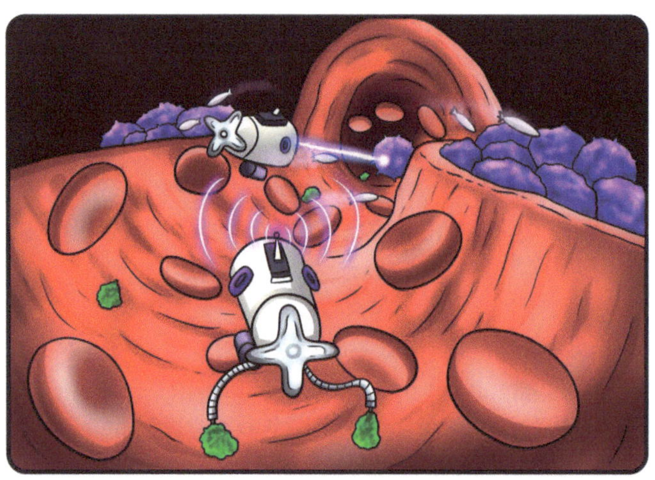

Bohrium could one day be integrated into nanorobotics, opening the door to precision micro-robotic surgery that feels like something from a sci-fi movie. In the future, tiny robotic systems powered by its unique properties might navigate delicate environments inside the human body with extraordinary accuracy. This could allow surgeons to target diseased tissue, repair damage, and minimize harm to surrounding cells, making advanced medical procedures safer, more effective, and far less invasive than many current techniques.

The Source of Bohrium

Element 107 (Bohrium) is synthesized by bombarding a target made of the post-transition metal element bismuth-209 with accelerated nuclei of the transition metal element chromium-54.

Bohrium is a strange, man-made metal. It is not found in nature. It only exists because scientists worked patiently, used advanced equipment, and followed clever ideas to create something that had never been seen before. Let's learn how Bohrium is made and about some of the people who helped bring it into existence.

Bohrium is a very heavy element. It belongs far out on the periodic table, in the group of elements that are difficult to make and even harder to study. Scientists do not find Bohrium in rocks, water, or air. Instead, they have to produce it in laboratories using special machines. Making it is not like mixing ingredients in a kitchen. It is more like setting up an extremely precise and powerful experiment where atoms are forced to collide. Scientists must plan every detail carefully, because if even one part goes wrong, the element will not form.

One common way to create Bohrium is through a nuclear reaction. In this process, scientists shoot a beam of one kind of atom at another kind of atom. You can think of it like trying to snap two tiny Lego bricks together, except the bricks are atoms and the force needed is much greater. The beam atom is usually a lighter element, and the target atom is much heavier. For Bohrium, researchers have used atoms such as chromium as the beam and heavy target atoms such as bismuth or similar materials. These atoms are sped up in a particle accelerator,

The Source of Bohrium (Continued)

which is a machine designed to push particles to very high speeds. When the beam hits the target with enough energy, the two nuclei may combine for a tiny moment and form a new atom. If that atom has the right number of protons, it is Bohrium.

This process is rare and difficult. Most collisions do not make Bohrium at all. Sometimes the atoms bounce apart. Sometimes they break into other pieces. Even when Bohrium is formed, it usually exists for only a very short time. It is highly unstable and quickly decays into smaller, more stable atoms. Because it disappears so fast, scientists cannot pick it up or hold it in a container. They have to detect it indirectly by observing the particles and radiation it leaves behind as it decays. Special detectors record these tiny signals, and scientists study them carefully to confirm that Bohrium really was created. This makes the discovery of Bohrium a bit like solving a puzzle with only a few brief clues.

The name Bohrium also has an interesting story. Scientists often name new elements after important people, places, or scientific ideas. Bohrium was named in honor of Niels Bohr, the famous physicist whose work helped explain the structure of atoms and the behavior of matter at very small scales. His ideas changed the way scientists understood the atom, so naming an element after him was a way to recognize his lasting influence on science.

Several people helped make Bohrium known to the world. One important team worked at the Gesellschaft für Schwerionenforschung, often called GSI, in Darmstadt, Germany. This research center specialized in experiments with heavy ions, which are atoms that have been stripped of electrons and then accelerated to very high speeds. In the late 1980s and early 1990s, scientists at GSI ran experiments designed to create superheavy elements, including Bohrium. These experiments took a great deal of skill, patience, and technical knowledge. The team had to choose the right atoms, set up the accelerator correctly, and watch for the smallest possible signs that the new element had appeared.

Two scientists often mentioned in the story of Bohrium are Peter Armbruster and Gottfried Münzenberg. Peter Armbruster was a leading figure in the Darmstadt research effort. He helped guide the experiments and the process of checking the results. Gottfried Münzenberg was another key scientist on the team, working closely with Armbruster to design experiments and interpret the data. Their collaboration was important because discoveries like Bohrium cannot be made by one person alone. They require many researchers who share ideas, double-check evidence, and work through many failed attempts before success finally comes.

Bohrium matters even though it lasts only a short time. Its creation teaches scientists how matter behaves at the extreme edge of the periodic table. It also shows the power of teamwork, careful measurement, and advanced technology. Every experiment with Bohrium helps scientists learn more about the forces that hold atomic nuclei together. Those lessons can be useful for understanding other heavy elements as well. In this way, Bohrium is not just a rare and unusual metal. It is also a symbol of human curiosity and persistence.

Bohrium reminds us that science often involves patience and imagination. Scientists build huge machines, design precise experiments, and keep searching even when success seems very unlikely. Thanks to that effort, Bohrium became real in the laboratory, and the people who helped create it earned an important place in the story of modern science.

Blaadlak The Goblin Fueled By Bohrium

Symbol: Bh Atomic Number: 107 Atomic Mass: 270

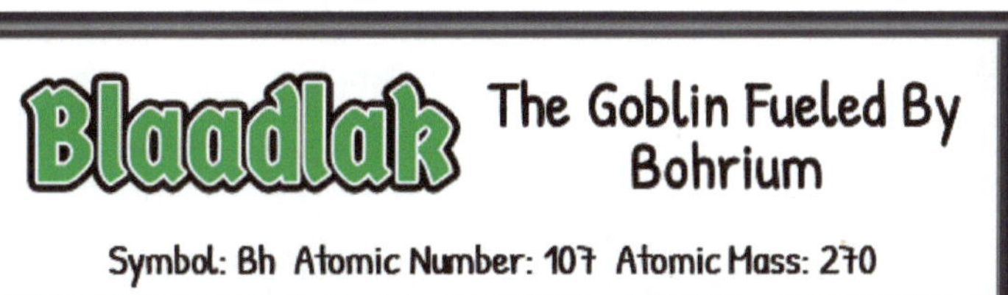

Bohrium resides in Group 7 Period 7 on the Periodic Table.
The atomic symbol is Bh. Its Atomic Number is 107. Its Atomic Mass is 270
As a member of the 6d series of transition metals and a heavy homologue of rhenium, it is located below rhenium (Re) in the table.

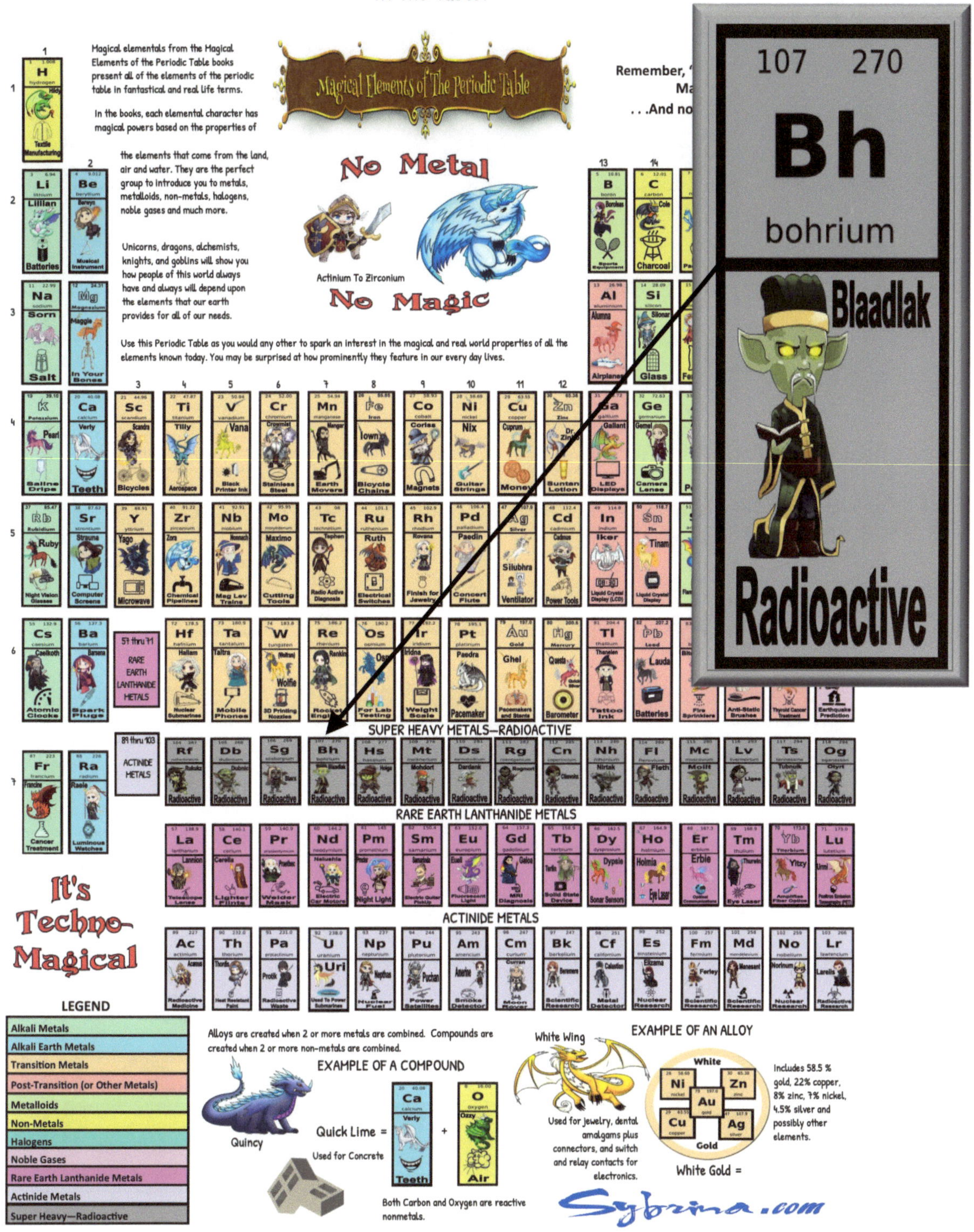

Magical Elements of The Periodic Table

Magical elementals from the Magical Elements of the Periodic Table books present all of the elements of the periodic table in fantastical and real life terms.

In the books, each elemental character has magical powers based on the properties of the elements that come from the land, air and water. They are the perfect group to introduce you to metals, metalloids, non-metals, halogens, noble gases and much more.

Unicorns, dragons, alchemists, knights, and goblins will show you how people of this world always have and always will depend upon the elements that our earth provides for all of our needs.

Use this Periodic Table as you would any other to spark an interest in the magical and real world properties of all the elements known today. You may be surprised at how prominently they feature in our every day lives.

No Metal
No Magic
Actinium To Zirconium

Remember, "No Metal— No Magic."
. . .And no technology.

H hydrogen Hildy Textile Manufacturing
He helium Hetha Balloons
Li lithium Lillian Batteries
Be beryllium Berwyn Musical Instrument
Na sodium Sorn Salt
Mg Magnesium Maggie In Your Bones
B boron Boroleas Sports Equipment
C carbon Cole Charcoal
N nitrogen Nitra Food Packaging
O oxygen Ozzy Air
F Fluorine Fleure Strong Bones and Teeth
Ne neon Jalan Advertising Signs
Al aluminium Alumna Airplanes
Si silicon Silonar Glass
P phosphorus Phova Fertilizer
S sulfur Xoe Matches
Cl chlorine Krystix Swimming Pools
Ar argon Areg Light Bulbs
K Potassium Pearl Saline Drips
Ca calcium Verly Teeth
Sc scandium Scandra Bicycles
Ti titanium Tilly Aerospace
V vanadium Vana Black Printer Ink
Cr chromium Crowmist Stainless Steel
Mn manganese Mangar Earth Movers
Fe Iron Iown Bicycle Chains
Co cobalt Coriss Magnets
Ni nickel Nix Guitar Strings
Cu copper Cuprum Money
Zn Zinc Dr. Zinko Suntan Lotion
Ga gallium Gallant LED Displays
Ge germanium Gemel Camera Lense
As arsenic Arkyn Poison
Se selenium Selenice Printers
Br bromine Brogach Photography Film
Kr krypton Krypto Detect Leaks
Rb Rubidium Ruby Night Vision Glasses
Sr strontium Strauna Computer Screens
Y yttrium Yago Microwave
Zr zirconium Zora Chemical Pipelines
Nb niobium Nonnach Mag Lev Trains
Mo molybdenum Maximo Cutting Tools
Tc technetium Tephen Radio Active Diagnosis
Ru ruthenium Ruth Electrical Switches
Rh rhodium Rovana Finish for Jewelry
Pd palladium Paedin Concert Flute
Ag Silver Silubhra Ventilator
Cd cadmium Cadmus Power Tools
In indium Iker Liquid Crystal Display (LCD)
Sn Tin Tinam Liquid Crystal Display
Sb antimony Antz Flame Resistant Fabric
Te tellurium Tellan Vulcanized Rubber
I Iodine Jody (Iodium) Cloud Seeding
Xe xenon Xena Used To Catch Speeders

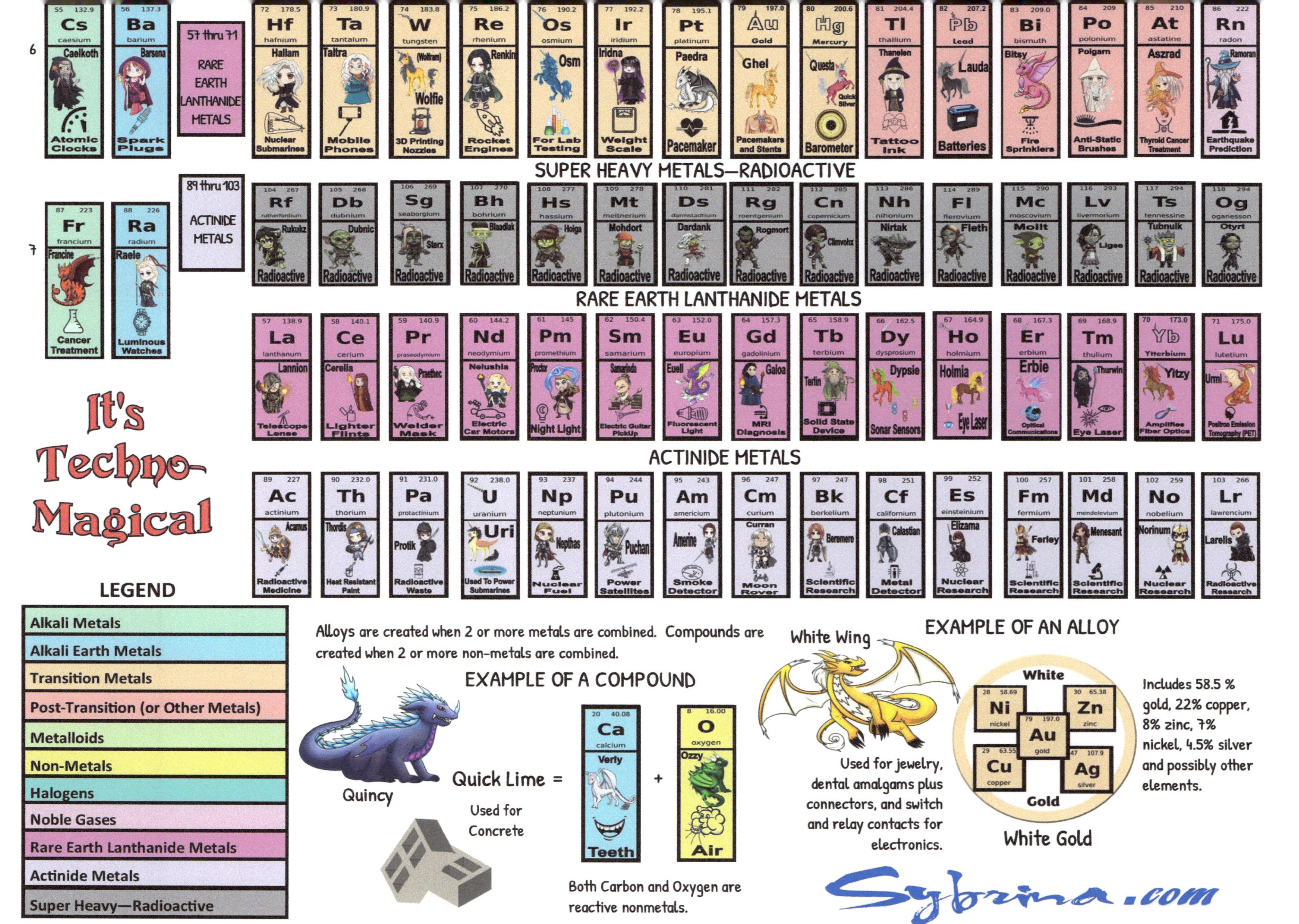
It's Techno-Magical

55 132.9 Cs caesium Caelkoth — Atomic Clocks
56 137.3 Ba barium Barsena — Spark Plugs
87 223 Fr francium Francine — Cancer Treatment
88 226 Ra radium Raele — Luminous Watches

57 thru 71 RARE EARTH LANTHANIDE METALS
89 thru 103 ACTINIDE METALS

72 178.5 Hf hafnium Hallam — Nuclear Submarines
73 180.9 Ta tantalum Taltra — Mobile Phones
74 183.8 W tungsten (Wolfram) Wolfie — 3D Printing Nozzles
75 186.2 Re rhenium Renkin — Rocket Engines
76 190.2 Os osmium Osm — For Lab Testing
77 192.2 Ir iridium Iridna — Weight Scale
78 195.1 Pt platinum Paedra — Pacemaker
79 197.0 Au Gold Ghel — Pacemakers and Stents
80 200.6 Hg Mercury Questa Quick Silver — Barometer
81 204.4 Tl thallium Thanelen — Tattoo Ink
82 207.2 Pb Lead Lauda — Batteries
83 209.0 Bi bismuth Bitsy — Fire Sprinklers
84 209 Po polonium Polgarn — Anti-Static Brushes
85 210 At astatine Aszrad — Thyroid Cancer Treatment
86 222 Rn radon Ramoran — Earthquake Prediction

SUPER HEAVY METALS—RADIOACTIVE

104 267 Rf rutherfordium Rukubz — Radioactive
105 268 Db dubnium Dubnic — Radioactive
106 269 Sg seaborgium Slerx — Radioactive
107 270 Bh bohrium Bloadlak — Radioactive
108 277 Hs hassium Hoiga — Radioactive
109 278 Mt meitnerium Mohdort — Radioactive
110 281 Ds darmstadtium Dardank — Radioactive
111 282 Rg roentgenium Rogmort — Radioactive
112 285 Cn copernicium Clinvotz — Radioactive
113 286 Nh nihonium Nirtak — Radioactive
114 289 Fl flerovium Fleth — Radioactive
115 290 Mc moscovium Molit — Radioactive
116 293 Lv livermorium Ligee — Radioactive
117 294 Ts tennessine Tubnulk — Radioactive
118 294 Og oganesson Otyrt — Radioactive

RARE EARTH LANTHANIDE METALS

57 138.9 La lanthanum Lannion — Telescope Lense
58 140.1 Ce cerium Cerelia — Lighter Flints
59 140.9 Pr praseodymium Praethec — Welder Mask
60 144.2 Nd neodymium Nelushia — Electric Car Motors
61 145 Pm promethium Proctor — Night Light
62 150.4 Sm samarium Sanariela — Electric Guitar PickUp
63 152.0 Eu europium Euell — Fluorescent Light
64 157.3 Gd gadolinium Galoa — MRI Diagnosis
65 158.9 Tb terbium Terlin — Solid State Device
66 162.5 Dy dysprosium Dypsie — Sonar Sensors
67 164.9 Ho holmium Holmia — Eye Laser
68 167.3 Er erbium Erbie — Optical Communications
69 168.9 Tm thulium Thurwin — Eye Laser
70 173.0 Yb Ytterbium Yitzy — Amplifies Fiber Optics
71 175.0 Lu lutetium Urmi — Positron Emission Tomography (PET)

ACTINIDE METALS

89 227 Ac actinium Acamus — Radioactive Medicine
90 232.0 Th thorium Thordis — Heat Resistant Paint
91 231.0 Pa protactinium Protik — Radioactive Waste
92 238.0 U uranium Uri — Used To Power Submarines
93 237 Np neptunium Nepthas — Nuclear Fuel
94 244 Pu plutonium Puchan — Power Satellites
95 243 Am americium Amerine — Smoke Detector
96 247 Cm curium Curran — Moon Rover
97 247 Bk berkelium Beremere — Scientific Research
98 251 Cf californium Calastian — Metal Detector
99 252 Es einsteinium Elizama — Nuclear Research
100 257 Fm fermium Ferley — Scientific Research
101 258 Md mendelevium Menesant — Scientific Research
102 259 No nobelium Norinum — Nuclear Research
103 266 Lr lawrencium Larelis — Radioactive Research

LEGEND
Alkali Metals
Alkali Earth Metals
Transition Metals
Post-Transition (or Other Metals)
Metalloids
Non-Metals
Halogens
Noble Gases
Rare Earth Lanthanide Metals
Actinide Metals
Super Heavy—Radioactive

Alloys are created when 2 or more metals are combined. Compounds are created when 2 or more non-metals are combined.

EXAMPLE OF A COMPOUND
Quincy
Quick Lime = Ca calcium + O oxygen
20 40.08 Ca calcium Verly — Teeth
8 16.00 O oxygen Ozzy — Air
Used for Concrete
Both Carbon and Oxygen are reactive nonmetals.

White Wing
EXAMPLE OF AN ALLOY
Used for jewelry, dental amalgams plus connectors, and switch and relay contacts for electronics.
White
28 58.69 Ni nickel
30 65.38 Zn zinc
79 197.0 Au gold
29 63.55 Cu copper
47 107.9 Ag silver
Gold
White Gold
Includes 58.5 % gold, 22% copper, 8% zinc, 7% nickel, 4.5% silver and possibly other elements.

Sybrina.com

All Of The Periodic Table Elements Listed Alphabetically

ACTINIUM—*AC*—89

ALUMINUM—*AL*—13

AMERICIUM—*AM*—95

ANTIMONY—*SB*—51

ARGON—*AR*—18

ARSENIC—*AS*—33

ASTATINE—*AT*—85

BARIUM—*BA*—56

BERKELIUM—*BK*—97

BERYLLIUM—*BE*—4

BISMUTH—*BI*—83

BOHRIUM—*BH*—107

BORON—*B*—5

BROMINE—*BR*—35

CADMIUM—*CD*—48

CALCIUM (Vital)—*CA*—20

CALIFORNIUM—*CF*—98

CARBON—*C*—6

CERIUM—*CE*—58

CESIUM—*CS*—55

CHLORINE (Keen)—*CL*—17

CHROMIUM—*CR*—24

COBALT—*CO*—27

COPERNICIUM—*CN*—112

COPPER—*CU*—29

CURIUM—*CM*—96

DARMSTADTIUM—*DS*—110

DUBNIUM—*DB*—105

DYSPROSIUM—*DY*—66

ERBIUM—*ER*—68

EINSTEINIUM—*ES*—99

EUROPIUM—*EU*—63

FERMIUM—*FM*—100

FLEROVIUM—*FL*—114

FLUORINE—*F*—9

FRANCIUM—*FR*—87

GADOLINIUM—*GD*—64

GALLIUM—*GA*—31

GERMANIUM—*GE*—32

GOLD—*AU*—79

HAFNIUM—*HF*—72

HASSIUM—*HS*—108

HELIUM—*HE*—2

HOLMIUM—*HO*—67

HYDROGEN—*H*—1

INDIUM—*IN*—49

IODINE (JODIUM)—*I*—53

IRIDIUM—*IR*—77

IRON—*FE*—26

KRYPTON—*KR*—36

LANTHANUM—*LA*—57

LAWRENCIUM—*LR*—103

LEAD—*PB*—82

LITHIUM—*LI*—3

LIVERMORIUM—*LV*—116

LUTETIUM (Unique)—*LU*—71

MAGNESIUM—*MG*—12

MANGANESE—*MN*—25

MEITNERIUM—*MT*—109

MENDELEVIUM—*MD*—101

MERCURY (QUICK SILVER)—*HG*—80

MOLYBDENUM—*MO*—42

MOSCOVIUM—*MC*—115

NEODYMIUM—*ND*—60

NEON (Jazzy)—*NE*—10

NEPTUNIUM—*NP*—93

NICKEL—*NI*—28

NIHONIUM—*NH*—113

NIOBIUM—*NB*—41

NITROGEN—*N*—7

NOBELIUM—*NO*—102

OGANESSON—*OG*—118

OSMIUM—*OS*—76

OXYGEN—*O*—8

PALLADIUM—*PD*—46

PHOSPHORUS—*P*—15

PLATINUM—*PT*—78

PLUTONIUM—*PU*—94

POLONIUM—*PO*—84

POTASSIUM—*K*—19

PRASEODYMIUM—*PR*—59

PROMETHIUM—*PM*—61

PROTACTINIUM—*PA*—91

RADIUM—*RA*—88

RADON—*RN*—86

RHENIUM—*RE*—75

RHODIUM—*RH*—45

ROENTGENIUM—*RG*—111

RUBIDIUM—*RB*—37

RUTHENIUM—*RU*—44

RUTHERFORDIUM—*RF*—104

SAMARIUM—*SM*—62

SCANDIUM—*SC*—21

SEABORGIUM—*SG*—106

SELENIUM—*SE*—34

SILICON—*SI*—14

SILVER—*AG*—47

SODIUM—*NA*—11

STRONTIUM—*SR*—38

SULFUR (Xanthous)—*S*—16

TANTALUM—*TA*—73

TECHNETIUM—*TC*—43

TELLURIUM—*TE*—52

TENNESSINE—*TS*—117

TERBIUM—*TB*—65

THALLIUM—*TI*—81

THORIUM—*TH*—90

THULIUM—*TM*—69

TIN—*SN*—50

TITANIUM—*TI*—22

TUNGSTEN—*W (WOLFRAM)*—74

URANIUM—*U*—92

VANADIUM—*V*—23

XENON—*XE*—54

YTTERBIUM—*YB*—70

YTTRIUM—*Y*—39

ZINC—*ZN*—30

ZIRCONIUM—*ZR*—40

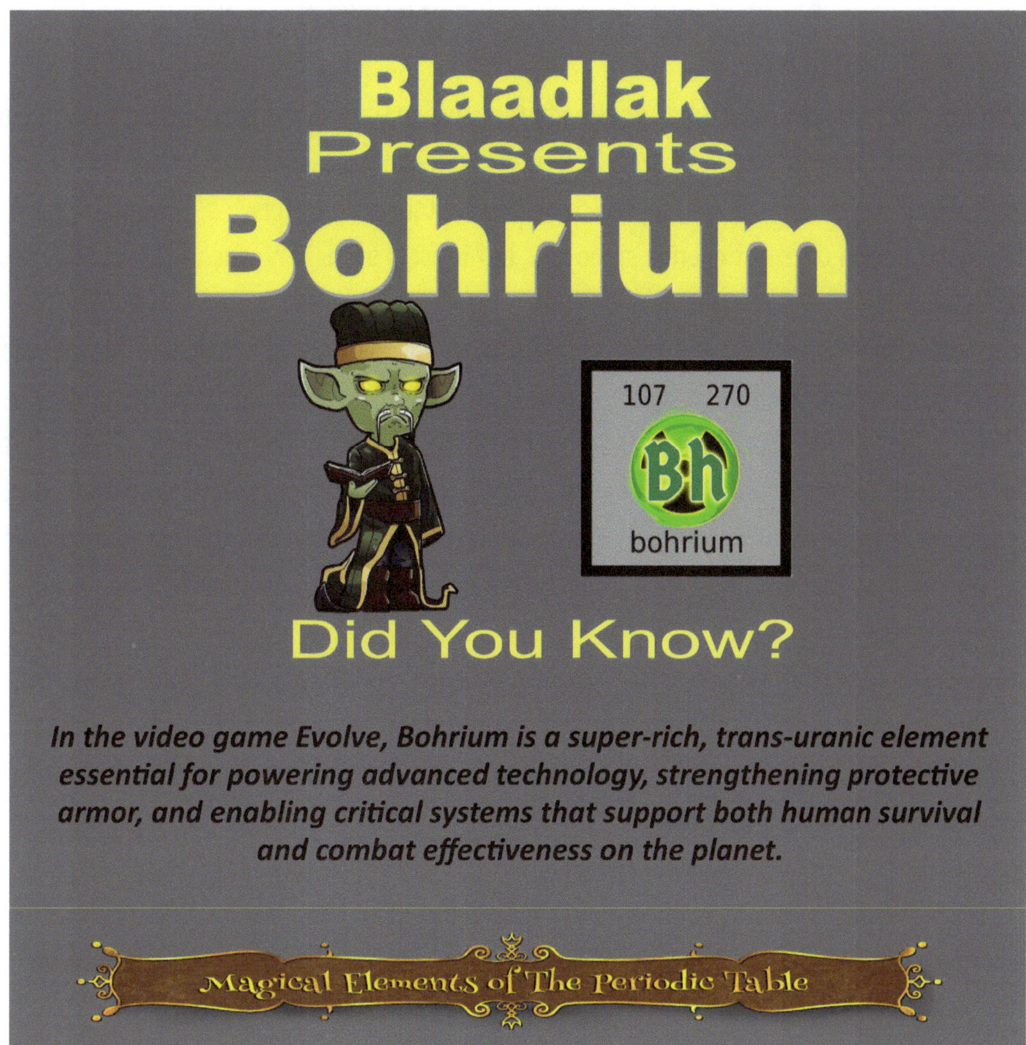

- Peter Armbruster, one of the people who helped create Bohrium, was an important nuclear physicist. Later in his career, he spent much of his time studying how particle accelerators could be used to change and break down dangerous nuclear waste that lasts a very long time. His work helped develop ideas for making radioactive materials less dangerous and less long-lasting.
- Bohrium is predicted to behave like Rhenium and Technetium on the periodic table because they are all in Group 7. A group on the periodic table is a vertical column of elements. Elements in the same group have similar chemical properties because they have the same number of valence electrons. This helps predict how they react and what compounds they form. The periodic table has 18 groups, numbered from 1 to 18.
- Bohrium was the first element made using a cold fusion process instead of a hot fusion process, where two nuclei are joined together. This method uses less energy, which helped scientists make the element in a more controlled way. Successfully making bohrium was an important step in nuclear chemistry and helped scientists learn more about how superheavy elements can be created.

Did You Know? (continued)

- Bohrium was named after Niels Bohr, who explained atoms as very small systems with a tiny, heavy center called the nucleus, with electrons moving around it in set energy levels. Electrons can move from one level to another by taking in or giving off energy, which helped explain atomic spectra. His model showed that matter at the tiniest scale does not change smoothly, but in fixed steps. This idea helped build modern quantum theory.
- Scientists generally think bohrium is not found in nature, on Earth or anywhere else in the universe. This is because bohrium is an unstable element made by people, not one that lasts for long. Some very heavy elements can be made in huge space events like supernova explosions or the merging of neutron stars. But bohrium is so radioactive that it breaks apart very quickly. Because of this, scientists believe it cannot stay around long enough to exist naturally anywhere in the universe. Even if some very heavy atoms were thrown out during a neutron star merger, it is very unlikely that any bohrium would survive long enough to be found.
- Scientists keep making bohrium (element 107) mainly to learn more about very heavy elements and to improve technology used in particle accelerators. Bohrium helps researchers study the chemical and physical behavior of superheavy elements and see how very large atoms act. The difficult steps needed to make and find bohrium also help improve particle accelerator technology, detection equipment, and methods used in nuclear research.
- Bohrium is usually thought to be just outside, or right on the edge of, the predicted "island of stability." It is in the transactinide area, and studies suggest it belongs to a group of "deformed shells" that link very unstable heavy atoms with the predicted island.

Structure Of Elements In The Periodic Table

Periodic tables are laid out in rows and columns.

Each element is placed in a specific location because of its atomic structure. Elements are arranged in Families.

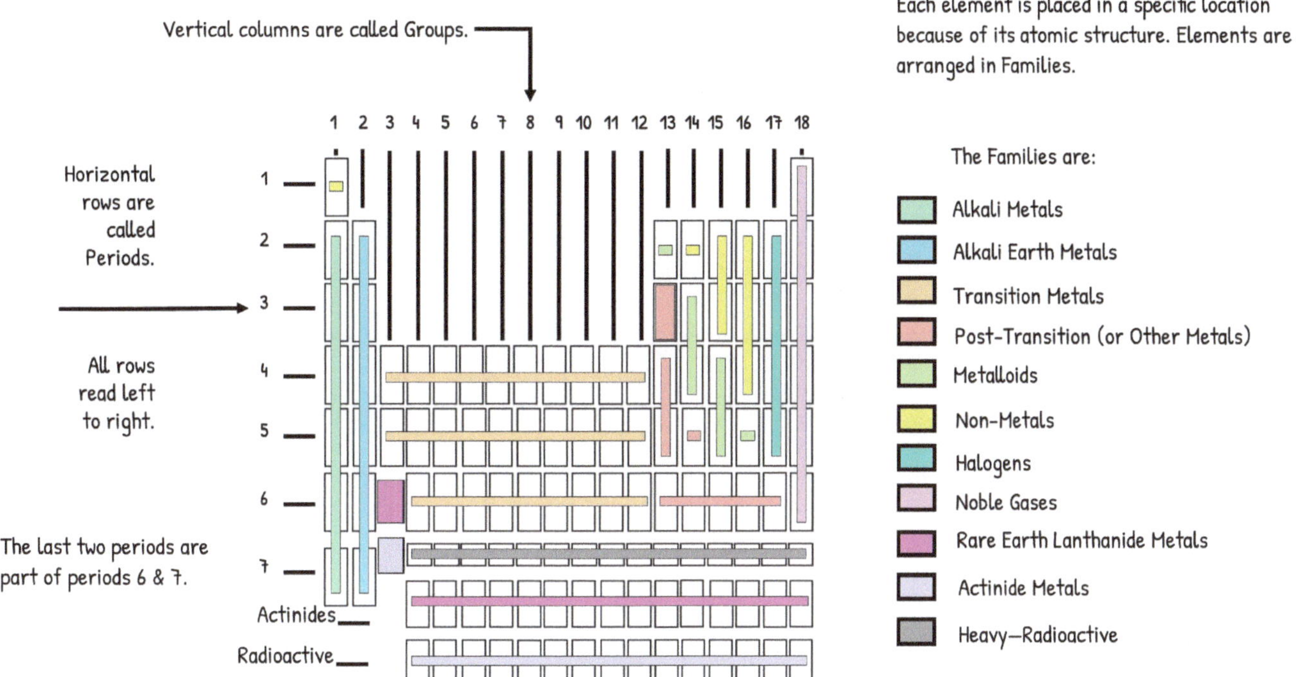

The term 'Element' is used to describe atoms with specific characteristics.
Every element in the first column or Group has 1 electron in the outer orbital (shell).
Every element in the second column (group two) has two electrons in the element's outer orbital.
The number designation of each Group represents the number of electrons in the element's outer orbital—
except for Group 18, Period 1—Helium. It only has 2 electrons.
Those electrons, called Valence Electrons, are what chemically bond with other elements.

Atomic Structure of Element: The atomic structure of an element refers to the arrangement of protons and neutrons in the nucleus of the atom, and the electrons in the electron cloud around the nucleus. Group 1, Period 1—Hydrogen is the only element that has no neutrons.

Bohrium (Bh) is located in Group 7 and Period 7 of the periodic table, positioning it as a d-block, transactinide element. It is a synthetic, highly radioactive transition metal, sitting below rhenium in the periodic table.

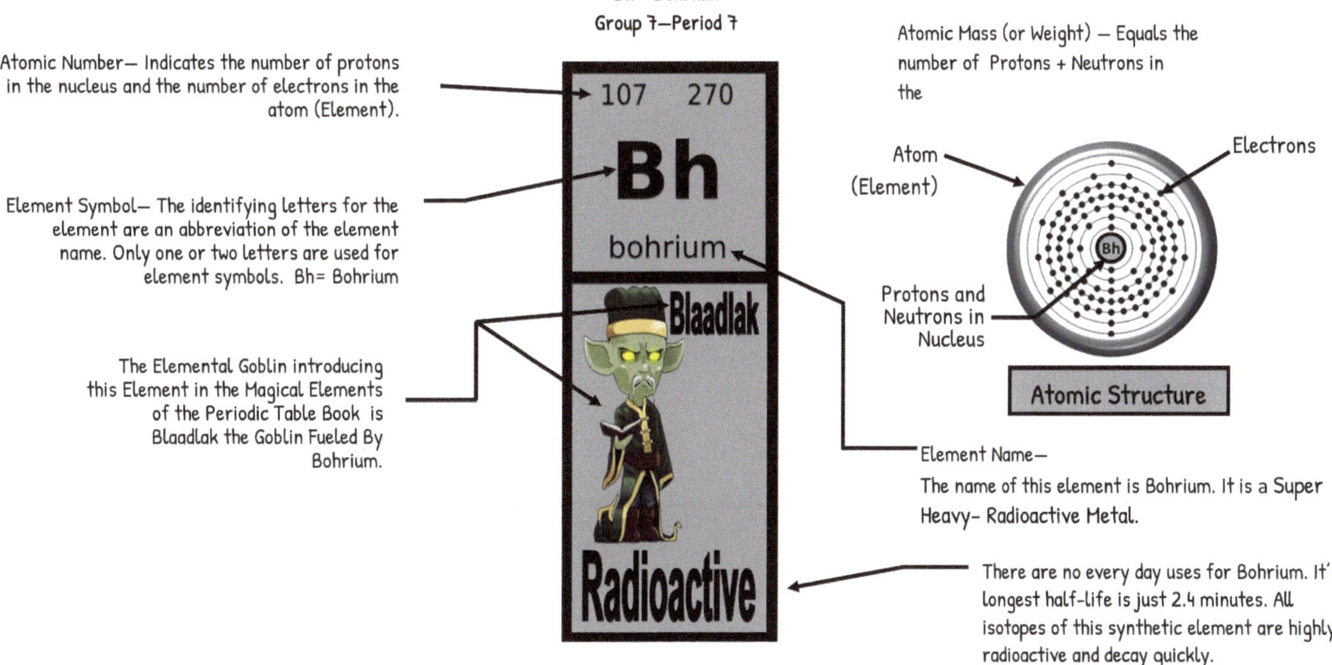

Atomic Number— Indicates the number of protons in the nucleus and the number of electrons in the atom (Element).

Element Symbol— The identifying letters for the element are an abbreviation of the element name. Only one or two letters are used for element symbols. Bh= Bohrium

The Elemental Goblin introducing this Element in the Magical Elements of the Periodic Table Book is Blaadlak the Goblin Fueled By Bohrium.

Atomic Mass (or Weight) — Equals the number of Protons + Neutrons in the

Element Name—
The name of this element is Bohrium. It is a Super Heavy- Radioactive Metal.

There are no every day uses for Bohrium. It's longest half-life is just 2.4 minutes. All isotopes of this synthetic element are highly radioactive and decay quickly.

Types of Elements On The Periodic Table

Alkali Metals—Some metals on the periodic table are soft and shiny. They are so soft that they can be cut with a knife! These metals are excited to give away electrons to elements in need, making them highly reactive. Ther electron transfer creates a compound known as a salt. Surprisingly, these metals are not found in nature alone; they must be extracted from other sources. Examples of these metals include lithium, sodium, potassium, rubidium, cesium, and francium.

Alkali Earth Metals—The elements in column 2 of the periodic table have 2 outer electrons in their shell. Ther makes them very active with nonmetals that need electrons to stay stable. When they react, they make something called a salt. They are often found in nature all by themselves, and they can even conduct electricity. The elements are beryllium, magnesium, calcium, strontium, barium, and radium.

Post-Transition (or other Metals)— Elements directly to the right of the transition metals. They are known as "poor metals: and are soft and brittle. These include aluminum, gallium, indium, tin, thallium, lead, bismuth, zinc, cadmium and mercury.

Transition Metal—The main metals are found in the middle and bottom rows of the periodic table. They look like metal, can conduct electricity, can bend and be shaped easily. The period 4 transition metals are scandium, titanium, vanadium, chromium, manganese, iron, cobalt, nickel, copper, and zinc. The period 5 transition metals are yttrium, zirconium, niobium, molybdenum, technetium, ruthenium, rhodium, palladium, silver, and cadmium. The period 6 transition metals are lanthanum, hafnium, tantalum, tungsten, rhenium, osmium, iridium, platinum, gold, and mercury. The period 7 transition metals are the naturally-occurring actinium, and the artificially produced elements rutherfordium, dubnium, seaborgium, bohrium, hassium, meitnerium, darmstadtium, and roentgenium.

Metalloids—The elements called metalloids are a mix of metals and nonmetals. They look like metals, but can't conduct electricity very well. They also break easily and act like nonmetals. These include boron, silicon, germanium, arsenic, antimony, tellurium, astatine, and polonium.

Non-Metals—These elements reside in columns 15-17, and can be gases, liquids, or solids. They don't conduct heat or electricity. The solids are brittle, and they have no metallic luster. They readily accept electrons from metals to form salts. These include nitrogen, oxygen, fluorine, chlorine, bromine, and iodine.

Halogens—Halogen chemicals are a special type of element. When they mix with metal, they become a kind of salt. Halogens are super reactive because they like to take an electron from metals. They can be found in column 17 of the element table. Some of them can be found in nature, but most are very dangerous and can hurt you if you touch them. They include fluorine, chlorine, bromine, iodine, and the radioactive elements astatine and tennessine.

Noble Gases—These elements reside in column 8. They are all odorless, colorless gases that are chemically very stable (inert). They don't generally form compounds by bonding with another element. These include helium, neon, argon, krypton, xenon, and radon.

Lanthanide Rare Earth Minerals—The Japanese call them "the seeds of technology." The US Department of Energy calls them "technology metals." These elements have atomic numbers 57-71. They are vital to industry. They can be added to metals to strengthen them to make alloys such as stainless steel, used to refine crude oil, and are crucial in producing technology—electronics, telecommunications, and metal devices to name a few. They are lanthanum, cerium, praseodymium, neodymium, promethium, samarium, europium, gadolinium, terbium, dysprosium, holmium, erbium, thulium,

Actinide Metals—Any of a series of chemically similar metallic elements with atomic numbers ranging from 89 (actinium) to 103 (lawrencium). All of these elements are radioactive, and two of the elements, uranium and plutonium, are used to generate nuclear energy. The lanthanides and actinides are sometimes called the inner transition metals, referring to their properties and position on the table. They are actinium, thorium, protactinium, uranium, neptunium, plutonium,

Super Heavy—Radioactive—Superheavy elements are those elements with a large number of protons in their nucleus. Elements with more than 92 protons are unstable; they decay to lighter nuclei with a characteristic half-life. They do not occur in large quantities (if at all) naturally on earth, and only exist briefly under highly controlled circumstances. They include lawrencium, rutherfordium, dubnium, seaborgium, bohrium, hassium, meitnerium, darmstadtium, roentgenium, copernicium, nihonium, flerovium, moscovium, livermorium, tennessine, and oganesson.

The "Island of Stability"

Normally, as elements become heavier, they become more unstable and decay instantly. The Island of Stability theory predicts a region of superheavy elements with "magic numbers" of protons and neutrons (like 114 protons and 184 neutrons) that would be significantly more stable, having longer half-lives (minutes, days, or even millions of years) than the fleeting microseconds of the highly unstable elements currently known beyond uranium. These special isotopes, though still radioactive, are expected to have longer lifetimes because their "doubly magic" nuclei are bound more strongly, potentially allowing scientists to study their unique chemical properties.

Elements 113 to 118, including Nihonium, Flerovium, Moscovium, Livermorium, Tennessine, and Oganesson, are currently the radioactive elements closest to the center of the theoretical Island of Stability, with isotopes showing slightly increased stability compared to their neighbors. Some predictions suggest that isotopes of these elements could have long enough half-lives to be studied extensively or even exist naturally, potentially making them useful for future technologies.

Scientists are actively synthesizing superheavy elements and measuring their properties, searching for evidence of this island, with recent discoveries showing promising, longer-than-expected half-lives that encourage the search. This finding could unlock entirely new types of matter, enable practical applications, and revolutionize chemical understanding.

SYNTHETIC RADIOACTIVE ELEMENTS & PERIODIC TABLE NEIGHBORS

Understanding chemical similarities through group relationships, despite nuclear instability

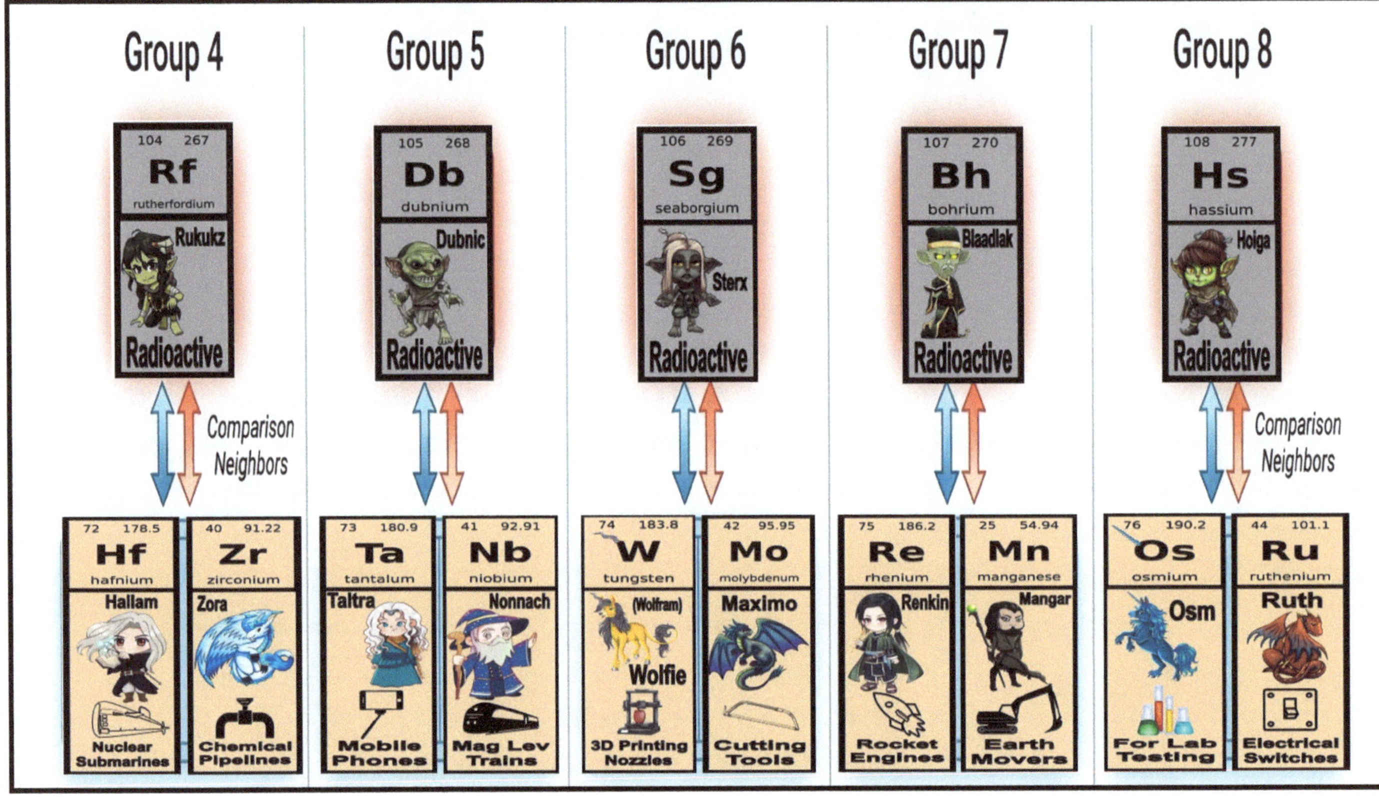

Why They Act Alike: The Valence Electron Connection

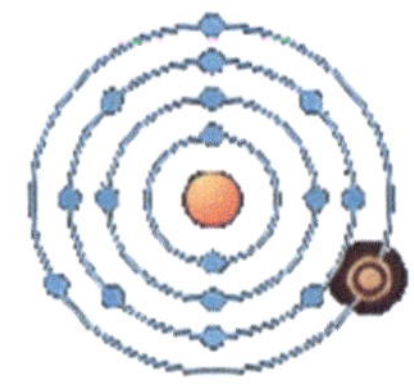

Elements in the same group share similar outer (valence) electrons, which primarily determine their chemical bonding and reactions. This structural similarity leads to comparable chemical behavor.

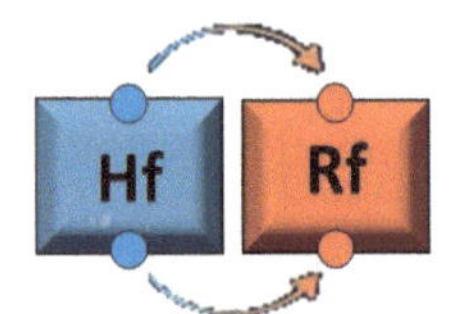

Examples of Chemical Similarity

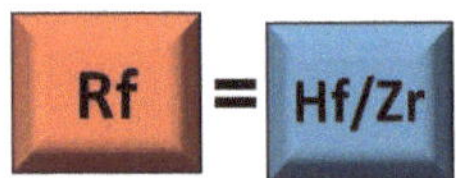

Rf behaves like Hf/Zr

(Similar formation of halides and oxides)

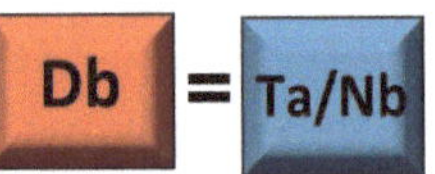

Db behaves like Ta/Nb

(Similar aquaeous chemistry, forming complex ions)

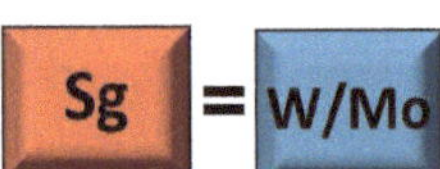

Sg behaves like W/Mo

(Similar behaviour in acidic solutions, forms stable oxides)

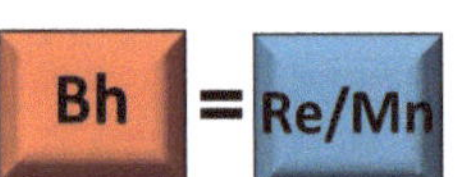

Bh behaves like Re/Mn

(Similar oxidation states, forms volatile oxides)

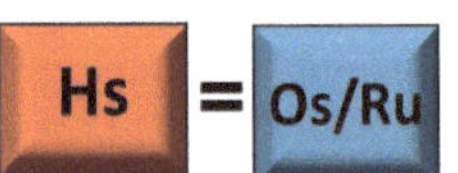

Hs behaves like Os/Ru

(Similar formation of tetroxides, high volatility)

Not Exact Copies

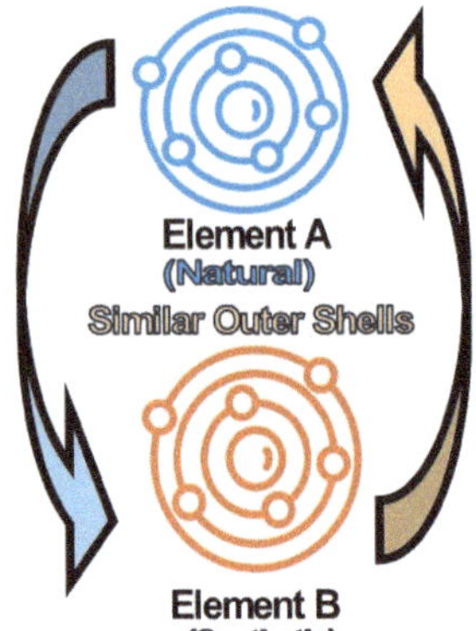

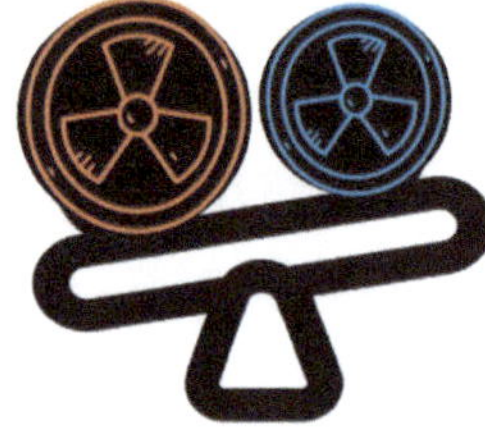

Differences Remain

While chemically similar, synthetic elements are:

Heavier Nuclei:
Significantly greater atomic mass and size.

Radioactive & Unstable

Short half-lives, decay quickly into other elements.

Relativistic Effects:
In superheavy elements, inner electrons move near light speed, subtly altering chemical properties compared to lighter neighbors.

Definitions

Atomic Structure of Element: The atomic structure of an element refers to the arrangement of protons and neutrons in the nucleus of the atom, and the electrons in the electron cloud around the nucleus.

Atomic Number: An element's atomic number refers to the number of protons it has in its nucleus. In a neutral atom the number of protons always equals the number of electrons.

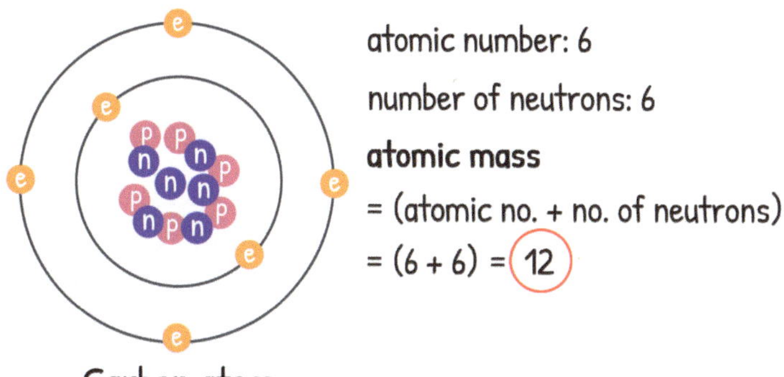

Atomic Weight (Mass) of Element: The atomic mass of an element is how heavy it is. It is made up of protons and neutrons that are in the middle of the element. Some elements have different versions with different amounts of neutrons, but they still have the same amount of protons. The atomic mass is the average weight of all these versions of the element.

Allotrope: Allotropes are different forms of an element that look and act different, but are made of the same stuff. Some elements have more than one form. For instance, carbon can be a shiny diamond or a gray pencil lead called graphite.

Isotope: Isotopes are different types of atoms that have the same parts, like protons and electrons, but they have a different number of neutrons. For example, the three most stable isotopes of hydrogen: protium (A = 1), deuterium (A = 2), and tritium (A = 3).

Crystalline Structure of Element: The crystalline structure of an element is how its atoms, ions, or molecules stick together in a pattern to make a cool crystal shape.

Ferrous and Non-Ferrous Metals: When we say ferrous metal, it means that iron is a big part of the metal. But if there's only a little bit of iron in the metal, we call it non-ferrous. The word "ferrous" comes from Latin and means iron, which is why iron's symbol is Fe.

Ductile Metals: These are capable of being made into long, thin wire or thread. Copper and Silver are ductile metals.

Malleable Metals: These can be hammered or rolled into thin sheets without cracking or breaking. Gold is malleable.

Ferromagnetic: Materials that are strongly attracted to a magnet. Such materials can be permanently magnetized. These include the elements iron, nickel and cobalt and their alloys, some alloys of rare-earth metals, and some naturally occurring minerals such as lodestone.

Magnetostriction—Ther is the term for a special thing that happens to magnetic materials. When these materials get turned into magnets, they also change their shape or size.

Paramagnetic: Slightly attracted to a magnetic field, but do not retain magnetic properties once the field is removed.

Diamagnetic: Slightly repelled by a magnetic field, but do not retain magnetic properties once the field is removed.

Electrical Properties: Conductor—a thing that lets electricity flow through it. Semi-conductor—a special material (usually silicon) that can conduct electricity, but not as well as metal. Insulator (non-conductor)—a material (usually glass) that stops electricity from flowing.

Reactive Gas: These gases are really good at reacting with stuff! They are called "sticky gases" because they can react to things like plastic and wet surfaces when they touch them. These are nitrogen, oxygen, hydrogen, carbon dioxide, fluorine, and chlorine.

Non-Reactive Gas: An inert gas is like a super shy gas that doesn't like to hang out with other chemicals. It doesn't make any new friends by reacting with them, so it doesn't form any chemical compounds. We also call these special gases "noble gases."

The Mysteries of Bohrium

Bohrium is an intriguing element on the periodic table, occupying atomic number 107 and carrying the name of Niels Bohr, the physicist whose ideas reshaped our understanding of atoms and quantum theory. Like many synthetic elements, Bohrium exists only because scientists managed to make it, and even then only in tiny amounts, for fleeting moments. It has no commercial use, and much about it remains uncertain, which only deepens its allure.

Its creation in 1976 felt something like a secret mission. Scientists at research laboratories tried to produce a new element by bombarding gold targets with chromium ions at extraordinary speeds. The process demanded precision and patience, more like threading a needle in a storm than conducting a routine experiment. Out of those collisions came evidence of Bohrium, a success that marked another step into the unknown.

Bohrium belongs to the transition metals and sits in period 7 among the heaviest members of the periodic table. Yet unlike stable metals that can be studied at length, Bohrium vanishes almost as soon as it appears. Its isotopes are highly unstable, and the most stable known form, Bohrium-270, lasts only about 61 seconds before decaying. That brief existence makes it difficult to observe directly, leaving most of its properties in the realm of theory and educated guesswork.

Even so, scientists look to its position in the periodic table for clues. Elements in the same general region often reveal patterns in how they behave, and Bohrium may share chemistry with other heavy transition metals. Researchers suspect it could one day help in high-energy physics, where even the smallest traces of behavior can illuminate the structure of matter itself. It might also offer insight into the strange conditions found in extreme environments, perhaps even those inside stars, though such possibilities remain distant and difficult to test.

What makes Bohrium especially compelling is not what it has done, but what it might still reveal. Because it is so short-lived and rare, it seems to hover at the edge of knowledge, just beyond reach. Scientists study it not for immediate practical value, but for the chance to understand how matter behaves when pushed to its limits. In that sense, Bohrium is less a tool than a question.

Its name gives the element a quiet elegance. Honoring Niels Bohr links it to the foundations of modern atomic theory and reminds us that every new discovery rests on earlier insight. Bohrium stands as a small, fleeting tribute to that legacy, a symbol of how science keeps moving forward by chasing what is barely visible.

For now, Bohrium remains a mystery. It does not power devices or shape industry, but it continues to matter because it points toward the unknown.

Meet Blaadlak, The Goblin
Fueled By Bohrium

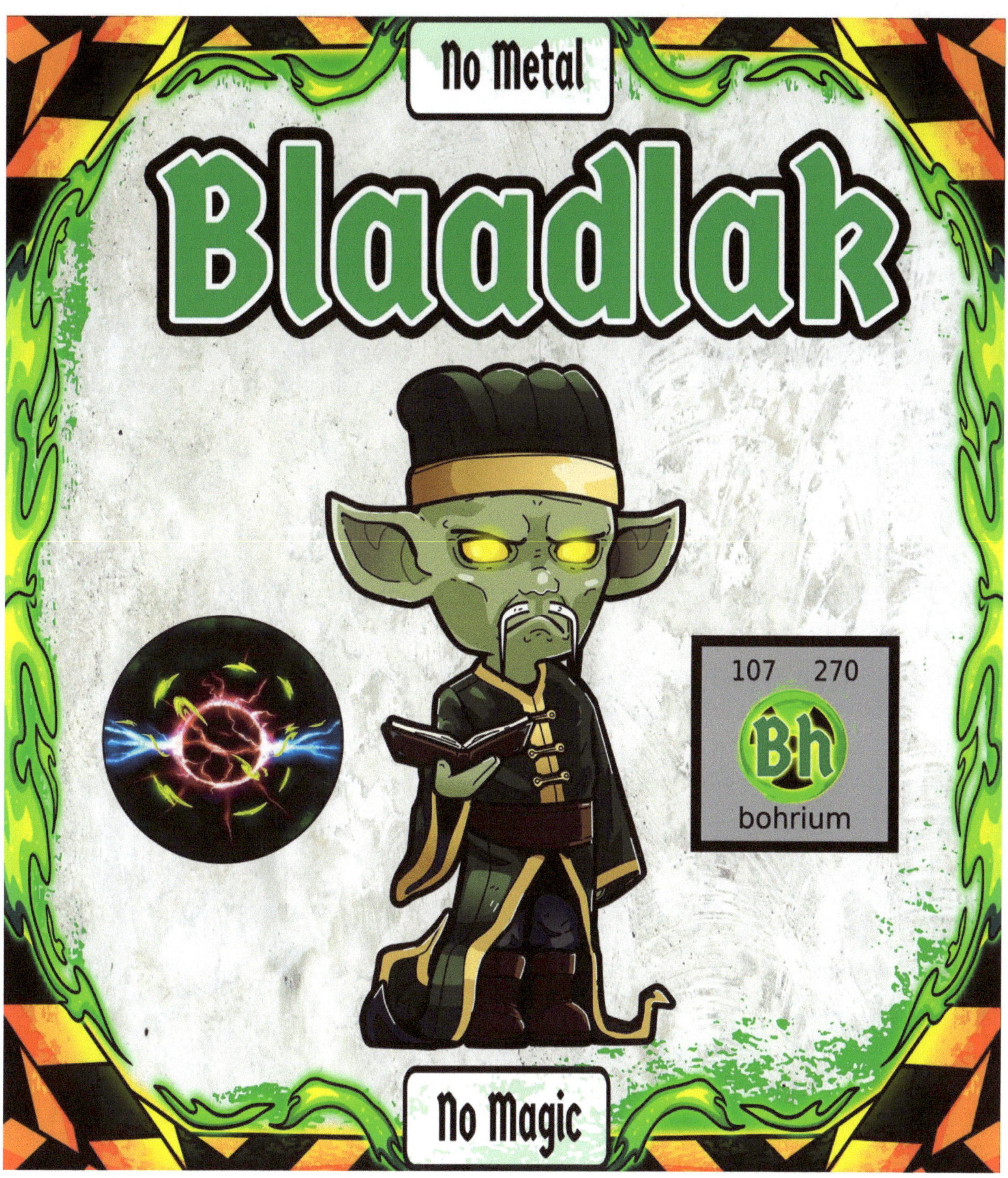

In the deep stacks beneath the ruined city of Vhorr, Blaadlak sat alone on a broken marble table, turning the pages of his weathered grimoire with careful hands. By the lantern's green glow, his mossy olive skin seemed to shift with every emotion, darkening when he frowned, brightening when he smiled. A tall ochre-wrapped headpiece crowned his bald skull, and his dark green velvet robe, trimmed in gold, made him look less like a common goblin than a scholar who had offended reality itself.

In his left hand was the grimoire. In his right, a small glass vial that pulsed with a sickly green light. Bohrium.

Few on the surface knew the name, and fewer still survived it. Blaadlak had found a distilled residue of the forbidden element hidden in a sealed chamber beneath the ruins, locked away by an empire long dead. The warning carved into the wall had been simple:

DO NOT WAKE WHAT THE METAL HAS SLEEPING IN IT.

Blaadlak had read it, smiled, and taken the vial anyway.

The first touch had nearly killed him. The second had changed him forever.

Now the bohrium lived in his blood like a caged star. It sharpened his mind, quickened his body, and surrounded him with a dangerous radiance that made the air shimmer and the bones ache. He was no longer merely clever or ambitious. He was something new: a goblin scholar with power enough to frighten armies.

And tonight, the ruined city had finally sent its answer.

Stone groaned overhead. Dust drifted from the ceiling. Far beyond the vault door, metal shrieked against metal.

"They found me," Blaadlak said, sounding almost pleased. "Late. Unprepared. Excellent."
The door burst inward.

Three mercenaries stumbled in first, shock-pikes raised. Behind them came a priest in hammered bronze armor and a white lacquer mask. Last entered Lady Seryth of the Ember Court, a human noblewoman in ash-colored gloves, elegant and cold, with the kind of smile that promised trouble. Her gaze went straight to the glowing vial.

"There you are," she said. "Our missing scholar."

Blaadlak gave her a precise little bow. "Your memory is generous."

"You stole from the Court."

"I borrowed from a tomb."

"You killed two of my men."

"I corrected their navigation."

The mercenaries shifted uneasily. Even from across the chamber, they could feel the bohrium's poison in the air. Lady Seryth did not flinch.

"Hand it over," she said.

Blaadlak looked at the vial, then at her. "You could ask more politely."

"We are past politeness."

"Then we are past your advantage."

He opened his grimoire and traced one blackened line with a finger. The page answered with a hiss of light. The glow in his skin flared. Bohrium stirred.

The chamber flooded with green radiance. The stone trembled. Blaadlak moved.

He crossed the room in a blur of velvet and static, too fast for the first mercenary to react. A shock-pike swung down and Blaadlak caught it bare-handed. The weapon sparked, failed, and died in his grip. Before the mercenary could scream, Blaadlak struck him in the throat and stepped away as the man collapsed.

The priest raised a warding charm. Blaadlak flicked a bone bead from his sleeve. It shattered against the charm in a burst of green sparks, and the priest recoiled as his protection broke.

Lady Seryth drew a slim runed blade. "You've changed."

"Adapted," Blaadlak said, grinning.

She lunged.

For a moment they moved like matched blades: her disciplined, precise, and lethal; him fast, wild, and unnervingly calm. Her sword cut through his shoulder, and the wound shone instead of bleeding. The pain made him hiss, but it only sharpened his focus.

"That," he said, "was rude."

He drew a cracked compass from his pocket and slammed it to the floor.

It exploded in a pulse of warped force. Lantern-light bent sideways. Stone chips lifted into the air. The mercenaries staggered. Lady Seryth was thrown back against a pillar, her blade spinning from her hand. Blaadlak did not waste the opening. He sprang past her, seized the priest by the mask, and pressed the vial to the bronze faceplate.

"Do you know what bohrium is?" he asked softly. "Not treasure. Not poison. A promise."

The priest trembled. "What promise?"

"That nothing stays stable forever."

He shoved the man aside.

Lady Seryth rose, blood at the corner of her mouth. "You can't control that power forever."

"No," Blaadlak said. "Only long enough."

He was not here simply to survive. He had come for what lay beneath the chamber: the machine the empire had built to use bohrium not as fuel, but as leverage. An ancient engine hidden under the city, meant to nudge fortune, shift alliances, and tilt the fate of nations by inches.

Not a weapon of destruction.

Something better. Something useful.

Blaadlak darted to the center of the vault and pressed the vial into a brass socket hidden beneath the floor. Glyphs ignited in a ring around him. The old machine woke with a low, hungry hum.

Lady Seryth shouted for the mercenaries to stop him, but it was too late.

The marble floor split open. Black metal pipes rose from below. Great gears turned for the first time in centuries. The chamber shook as if a buried heart had begun to beat again.
Green light flooded the ruins.

For an instant, every candle in the lower stacks burned emerald. The mercenaries fled. The priest crawled away in terror. Lady Seryth stared at Blaadlak as though she had finally understood what he was: not a thief, not a madman, but a mind that had chosen danger and taught it discipline.

Then the ceiling cracked.

The machine was waking too violently. Dust rained down. A deep tremor rolled through the chamber.

Blaadlak felt the bohrium in his veins surge and howl, threatening to tear him apart from the inside. For the first time, his smile faded.

He understood at once what had gone wrong. The machine was not merely drawing on the bohrium. It was trying to draw on him as well.

Lady Seryth, seeing the strain in his face, made a choice. She could have fled. Instead, she stepped toward the control ring and seized one of the brass levers before the collapse could spread.

"What are you doing?" Blaadlak shouted.

"Saving my own city," she snapped. "Try not to sound surprised."

Together, against all reason, they worked. Blaadlak used the grimoire to stabilize the runes while Seryth forced the ancient gears into alignment. The machine shuddered, groaned, and finally settled into a steady rhythm. The ceiling's rumbling began to slow, then stop.

Silence returned in a trembling wave.

The green glow dimmed to a soft pulse beneath the floor.

Blaadlak leaned against the table, breathing hard, one hand pressed to his shoulder. Lady Seryth stood across from him, dust on her gloves and a new wariness in her eyes.

"You could have let it collapse," she said.

"And lose my investment?" he replied.

Despite herself, she almost smiled.

The machine remained alive beneath the ruins, but contained. Not a conquest engine, not a doomsday relic, but a buried source of power that could be studied, negotiated with, and—if Blaadlak had his way— used with care and cunning.

Lady Seryth looked at the glowing floor. "The Ember Court will want this."

"Of course it will."

"And so will every thief, mage, and fool in three kingdoms."

"Yes," Blaadlak said. "Which means I'll need allies."

She studied him for a long moment. Then she extended one ash-gloved hand.

"Careful, scholar," she said. "You may regret trusting me."

Blaadlak took her hand with a slight bow. "Regret is for people with fewer options."

Above them, Vhorr remained broken and silent, but beneath it the old machine beat on, no longer sleeping, no longer lost.

Blaadlak smiled—not with madness, but with purpose.

In the dark beneath the city, he had not merely survived the power of bohrium. He had claimed responsibility for it. And in doing so, he had done the one thing no empire had managed before: He had made chaos useful.

Enjoy This Coloring Page Featuring

Blaadlak The Radioactive Goblin Fueled By Bohrium

Sample Page From Magical Elements of the Periodic Table

Presented By The Radioactive Goblins

MEET THE RADIOACTIVE GOBLINS

Blaadlak—Bohrium

Climvohx—Copernicium

Dardank—Darmstadtium

Dubnic—Dubnium

Fleth—Flerovium

Hoiga—Hassium

Ligee—Livermorium

Mohdort—Meitnerium

Moilt—Moscovium

Nirtak—Nihonium

Otyrt—Oganesson

Rogmort-Roentgenium

Rukukz—Rutherfordium

Sterx—Seaborgium

Tubnulk—Tennessine

THEIR MAGICAL POWERS ARE FUELED BY RADIOACTIVE ELEMENTS FROM THE PERIODIC TABLE

Create Your Own Magical Goblin Elemental

Blaadlak — The Goblin Fueled By Bohrium

Symbol: Bh Atomic Number: 107 Atomic Mass: 270

Bh

107 270
Bh
bohrium

Magical Elemental Symbol

Bohrium is a Super Heavy Radioactive Metal

Synthesized by bombarding Bismuth-209 with Chromium-54 in nuclear reactors

107 270
Bh
bohrium
Bh
Radioactive
Blaadlak

Bohrium Periodic Symbol

Blaadlak's Magical Abilities

Blaadlak craves knowledge and power. He excels in alchemy and chemistry, crafting potent potions and poisons.

Atomic Structure

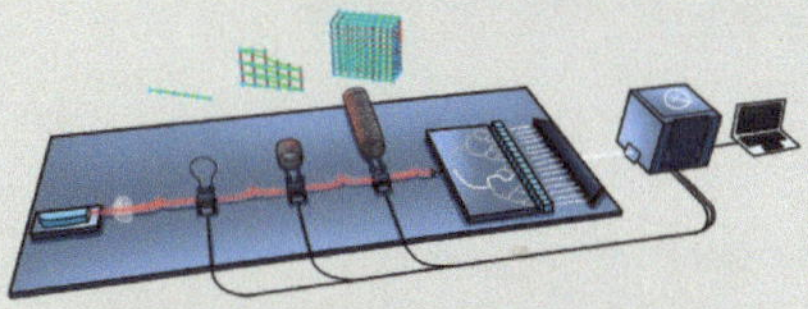

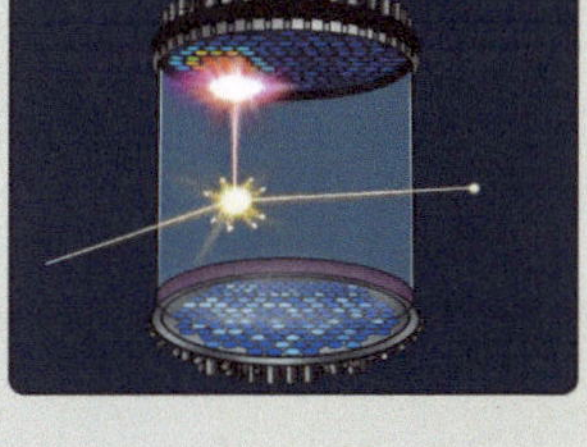

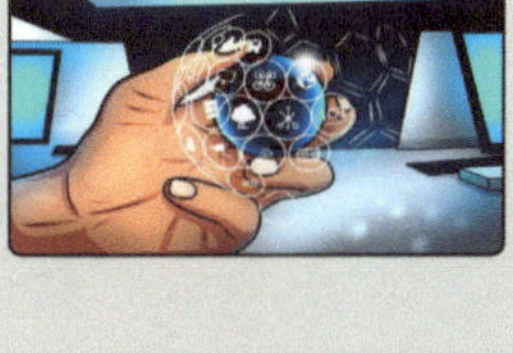

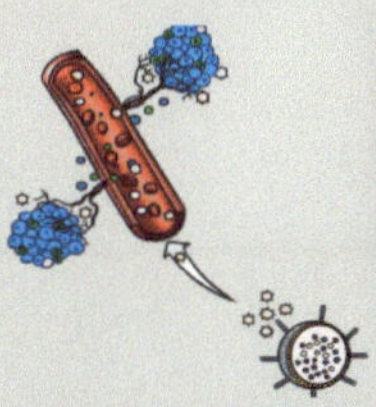

Magical Elements of The Periodic Table

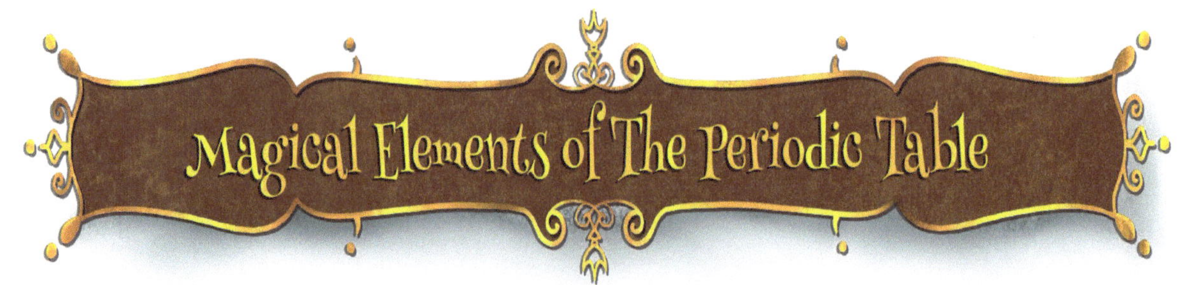

Students may either use a program like power point to cut and paste clip art into a Magical Goblin Elemental Blank or, if they wish, they may draw everything themselves.

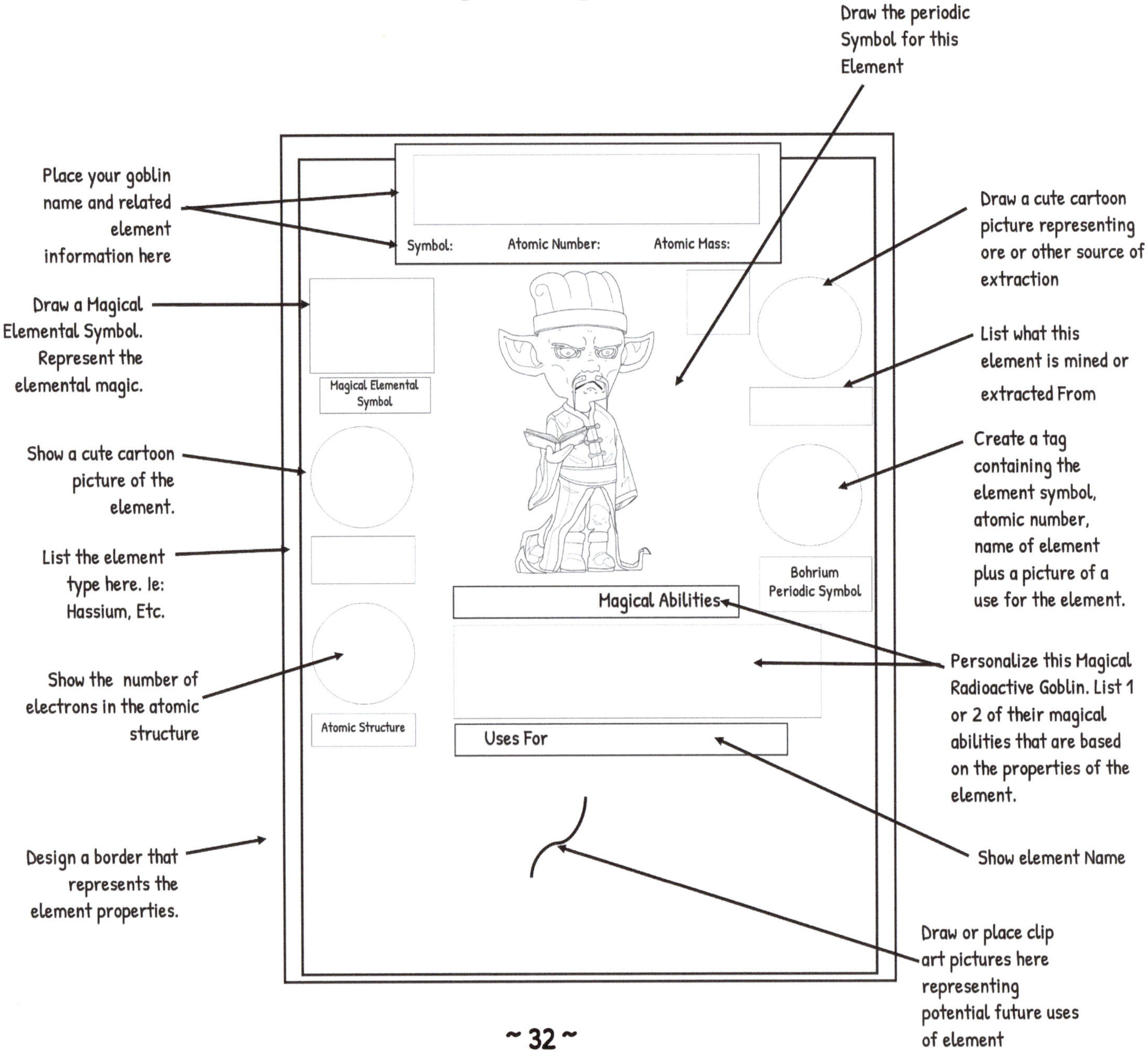

Draw the periodic Symbol for this Element
Place your goblin name and related element information here
Symbol:
Atomic Number:
Atomic Mass:
Draw a cute cartoon picture representing ore or other source of extraction
Draw a Magical Elemental Symbol. Represent the elemental magic.
Magical Elemental Symbol
List what this element is mined or extracted From
Show a cute cartoon picture of the element.
Create a tag containing the element symbol, atomic number, name of element plus a picture of a use for the element.
List the element type here. Ie: Hassium, Etc.
Bohrium Periodic Symbol
Magical Abilities
Show the number of electrons in the atomic structure
Atomic Structure
Personalize this Magical Radioactive Goblin. List 1 or 2 of their magical abilities that are based on the properties of the element.
Uses For
Design a border that represents the element properties.
Show element Name
Draw or place clip art pictures here representing potential future uses of element

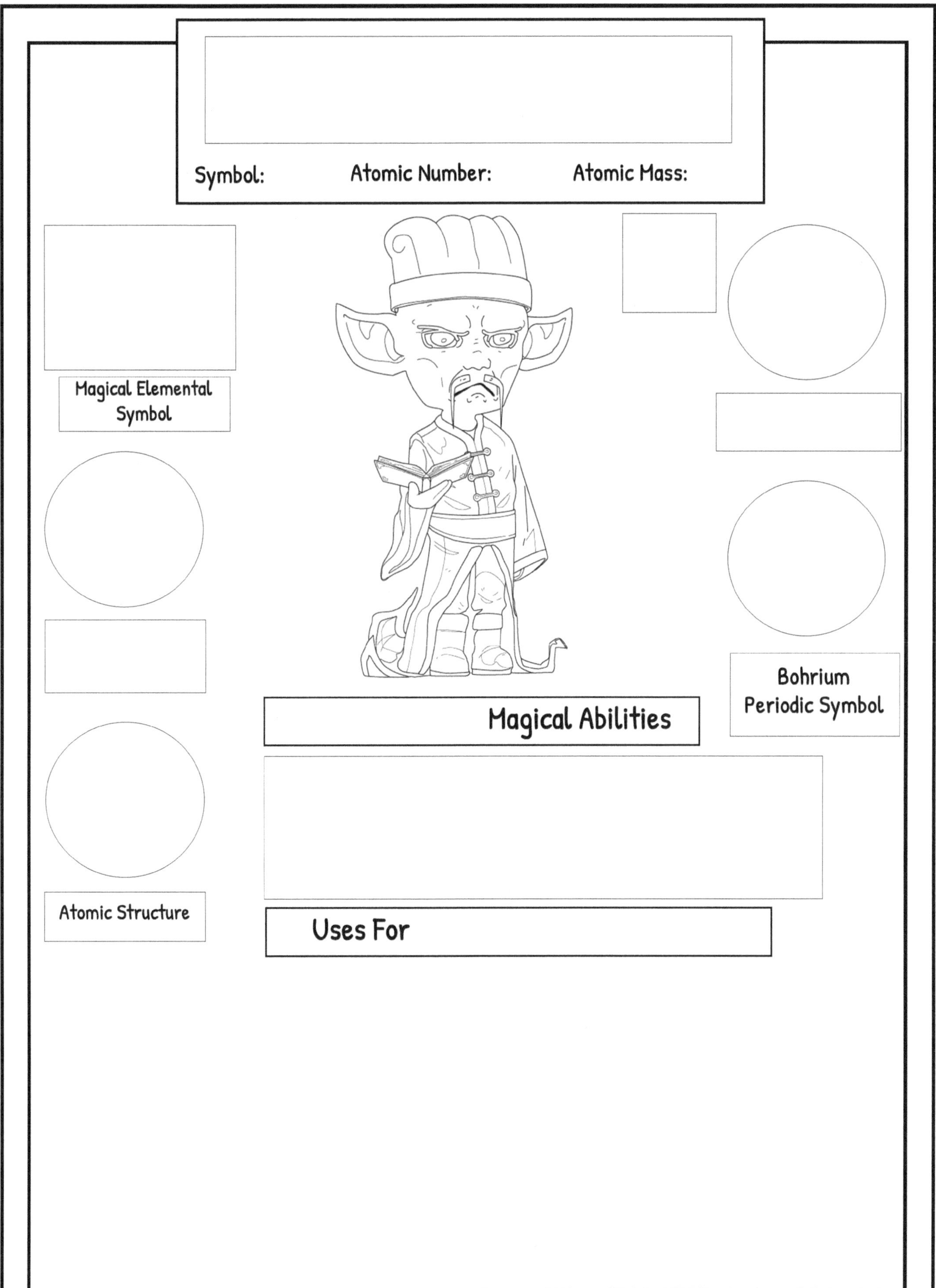

Symbol:　　　　Atomic Number:　　　　Atomic Mass:

Magical Elemental
Symbol

Magical Abilities

Bohrium
Periodic Symbol

Atomic Structure

Uses For

Magical Goblin Elemental Research Sheet

Before starting your Magical Goblin Elemental graphics page, do some research on your chosen element.

Name of Magical Goblin:	
Goblin's Magic Power Based on the Element's Properties:	
Magical Elemental Symbol:	
Element Name:	
Element Symbol:	
Atomic Number:	
Atomic Mass:	
What year and where was this Element discovered?	
Who discovered this Element?	
Element Group:	
Element Period:	
Element Family Name:	
State of Element At Room Temperature:	
How is the Element Created?	
Is Element Magnetic?	
Does Element Conduct Electricity?	
If this Element were metal, what would is it expected to look like?	
What is shortest life span of this Element?	
What is the longest life span of this Element?	
What element on periodic table is this one expected to behave most like?	
Name one real use for this Element:	
Invent a SciFi use for this Element:	
Interesting and Fun Facts about the discovery and creation of this element:	

Magical Unicorn Elemental Research Sheet

Before starting your Magical Unicorn Elemental graphics page, do some research on your chosen element.

Name of Magical Unicorn:	Ghel The Gold Horn Unicorn
Unicorn's Magic Power Based on the Element's Properties:	Ghel can see past, present and future. She is empathic and can sympathize with the feelings of other. They say she has a heart of gold.
Magical Herd Crest Symbol:	An open heart with a Celtic Trinity Knot.
Element Name:	Gold
Element Symbol:	Au— Comes from Aurum which is the Latin word for Gold.
Atomic Number:	79
Atomic Mass:	196.97
What year and where was this Element discovered?	Around 4,600 BCE in Bulgaria
Who discovered this Element?	Unknown
Element Group:	11 on Periodic TAble
Element Period:	6 on Periodic TAble
Element Family Name:	Gold is a Noble Transition Metal
State of Element At Room Temperature:	Solid
What is Element Mined or Extracted From?	Quartz Veins. It is also found in gravel in streams.
Is Element Magnetic?	It is Diamagnetic. It's only weakly magnetized when placed in a magnetic field.
Does Element Conduct Electricity?	Gold is a great electrical conductor used in printed circuitry of computers.
Where is the Element commonly found in Nature?	One of the largest deposits is found in the United States in Arkansas.
What is 1 alloy of the Element? How used?	White gold is an alloy of gold, palladium, nickel and zinc.
What is 1 compound of the Element? How used?	Gold Phosphide is a semiconductor used in high power, high frequency applications and in laser diodes.
Name the most common use for this Element:	Jewelry
Name a little known use for this Element:	Acupuncture needles
Name one more use for this Element:	Gold is used in airbags in cars.
Interesting and Fun Facts:	Gold was used in ancient Egypt to fill decayed teeth. Gold thread is incorporated in astronaut spacesuits to protect them from the heat of the sun.

Write a paragraph below to describe your magical goblin elemental. Based on the information obtained from research of your chosen element, how did you determine your goblin's name? What are your goblin's magic powers? What are their likes/dislikes, strengths/weaknesses, personality traits? What color are your goblin's armor and clothing and why did you pick that/those colors? What is your goblin's Magical Elemental Symbol?

Magical Unicorn Elemental Sample Description

Ghel The Gold-Horned Unicorn

This magical unicorn has a golden horn and hooves that glow like the sun. Her hide is honey-gold and her flowing mane and tail are golden-blonde.

Ghel is a member of the Metal Horn Unicorn Tribe from Unimaise. Gold is linked to the heart chakra because it holds a warm energy that brings soothing vibrations to the body to aid in the healing process. Ghel's Magical Herd Crest symbol is an open heart with a Celtic Trinity knot which indicates that she can see past, present and future. She is empathic and can sympathize with the feelings of others.

Being empathic does not make Ghel a weakling. She is brave with strong opinions and is a true champion to those she loves. She is known as the unicorn with the "heart of gold". When she places her horn on the heart of another, she senses their future.

The name Ghel is an Indo-European word which means yellow. The word "gold" most likely has its origins in the word "Ghel".

Gold is known to possess spiritual powers that bring happiness, peace, stability and luck to those who wear it. Scientists say that all the gold in the world comes from the collision of neutron stars.

Though most other magical unicorn elementals get along well with the gold horn unicorn herd; Ghel, and others like her, must be very careful around the Quick Silver Herd - as gold dissolves in mercury.

Do Your Middle Graders Want To Know More From The Magical Elementals About The Periodic Table?

Get the accompanying books in print at all online book stores. Get the books and accompanying activities at MagicalPTElements.

Available Now

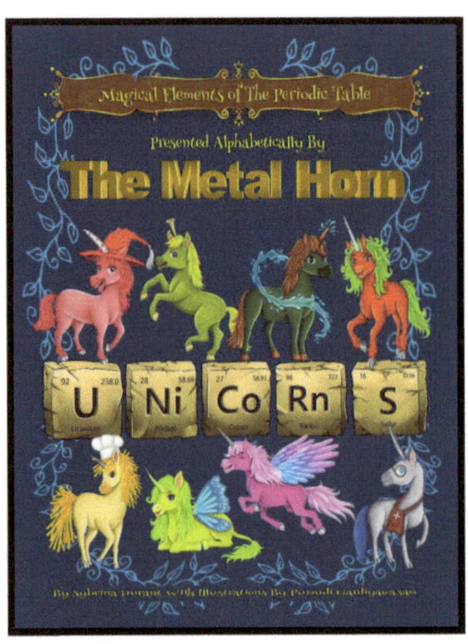

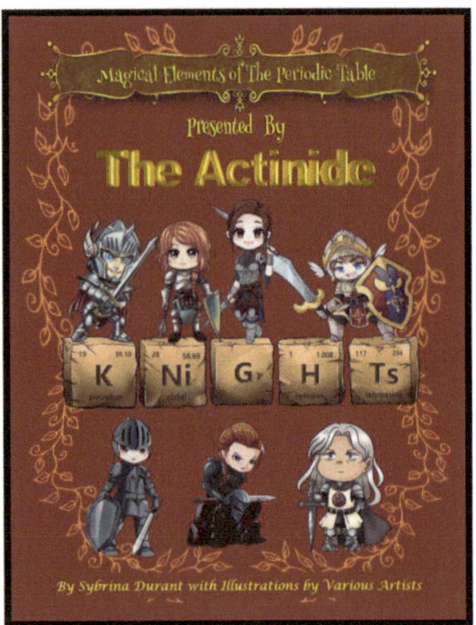

Get These Trading Cards Sets Featuring The Magical Elementals Representing The Periodic Table Elements

at https://bit.ly/40oEUBr

Unicorns, Dragons, Wizards, Knights, or Goblins?

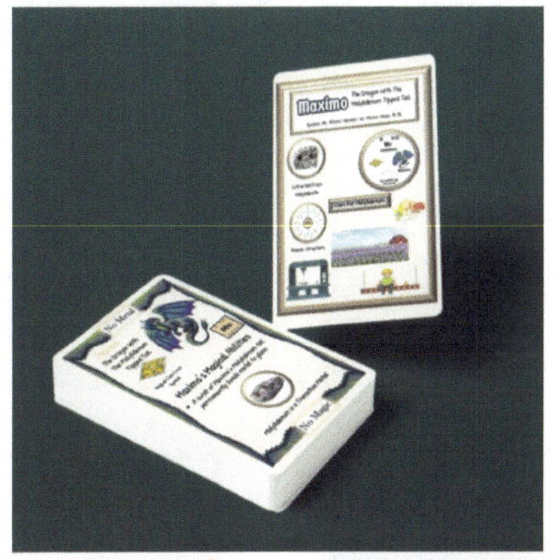

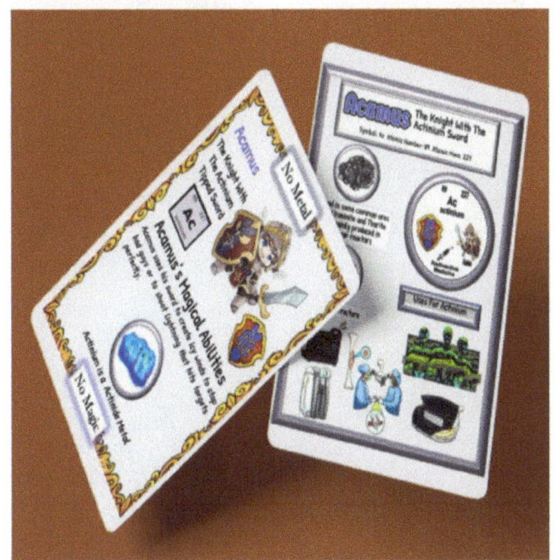
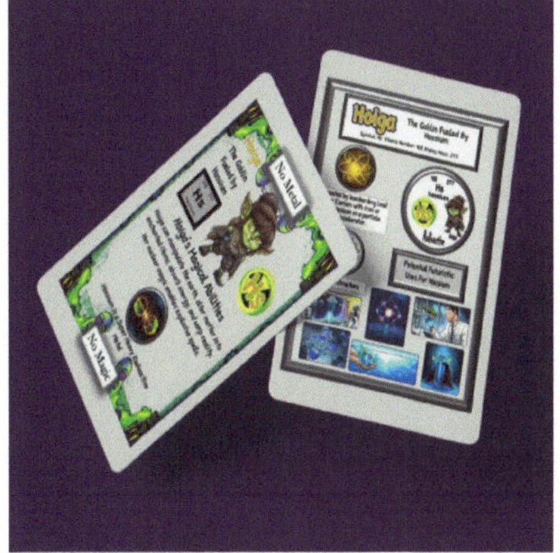

Collect Them All

These magical elementals are ready to help make learning the periodic table more fun. Get all books and related activities today.

Learn More About all of the Periodic Table Elementals. Get all of the "No Metal No Magic" Books Featuring Individual Elements at MagicalPTElements.com

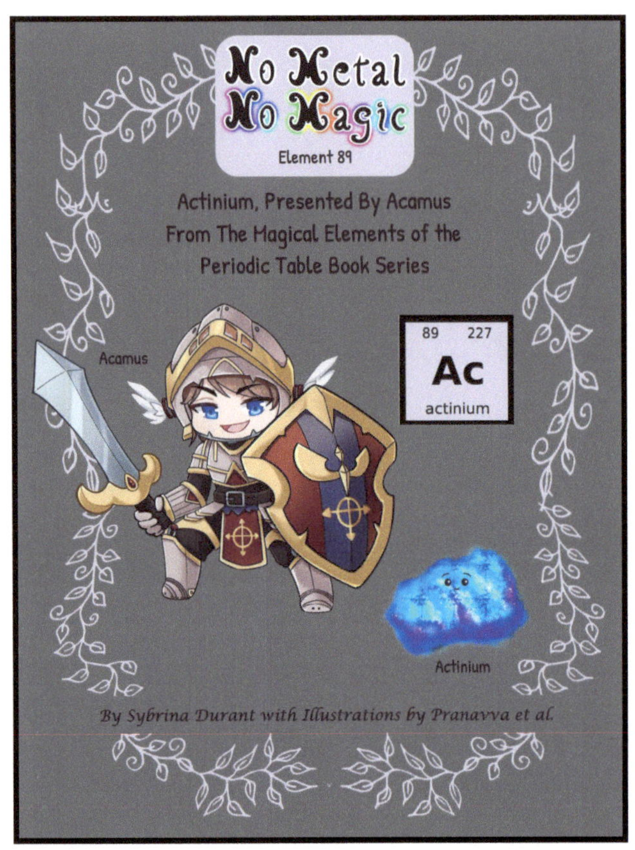

Get These Fun Elemental Periodic Table Activities at

MagicalPTElements.

Unicorn Periodic Table Bingo—Comes with 32 unique Bingo cards. Magical Elementals Bingo comes with 36.

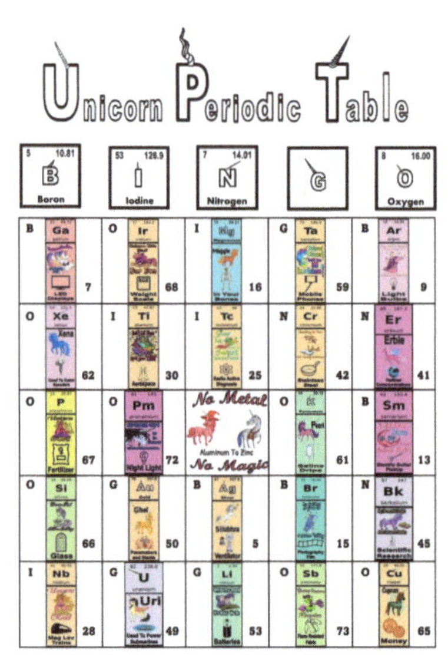

Magical Elemental Game Cards—Makes great prizes. Fun to trade, too.

1
2
3

Unicorn **H**orn

plus

Alphabet

CLIP ART FOR YOUR GRAPHICS

A
B
C

Also browse activities at
https://www.magicalPTelements.com
for all kinds of printable downloads to make learning fun.

Printable Magical Elemental Activity Downloads

Fun Way For Students To Learn The Elements Of The Periodic Table

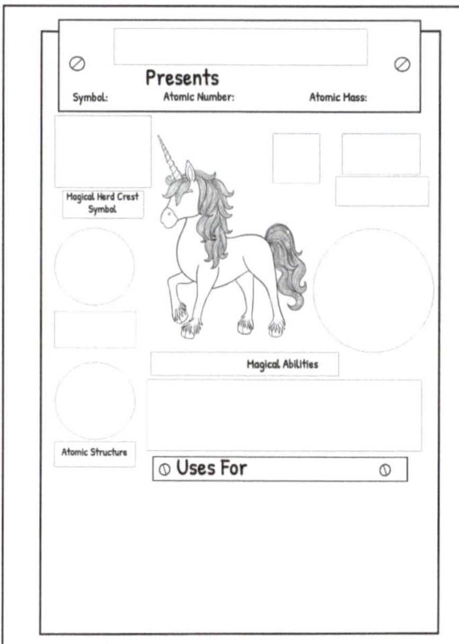

Blank Unicorn Element Card

Sample Unicorn Element Card

Magical Unicorn Elemental Research Sheet

Before starting your Magical Unicorn Elemental graphics page, do some research on your chosen element.

Name of Magical Unicorn:

Unicorn's Magic Power Based on the Element's Properties:

Element Name:

Element Symbol:

Number of Protons:

Number of Neutrons:

Number of Electrons:

Element Group:

Element Period:

Element Family Name:

Element Type:

State of Element At Room Temperature:

What is Element Mined or Extracted From?

Is Element Magnetic?

Does Element Conduct Electricity?

Where is the Element commonly found in Nature?

What are 2 alloys of the Element?

What are 2 compounds of the Element?

Name the most common use for this Element:

Name a little known use for this Element:

Name one more use for this Element:

What year and where was this Element discovered?

Who discovered this Element?

Blank Research Sheet

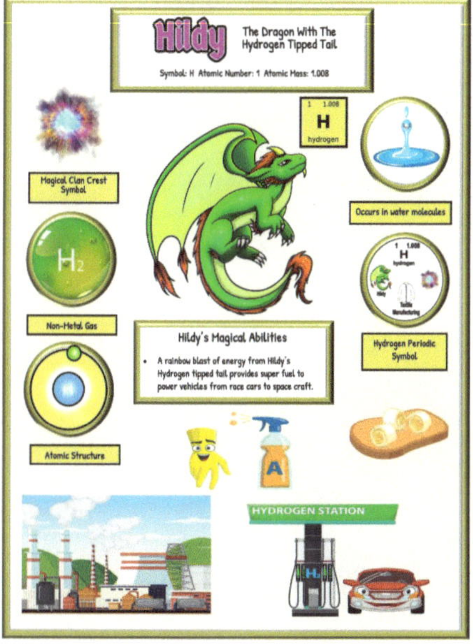

Sample Dragon Element Card

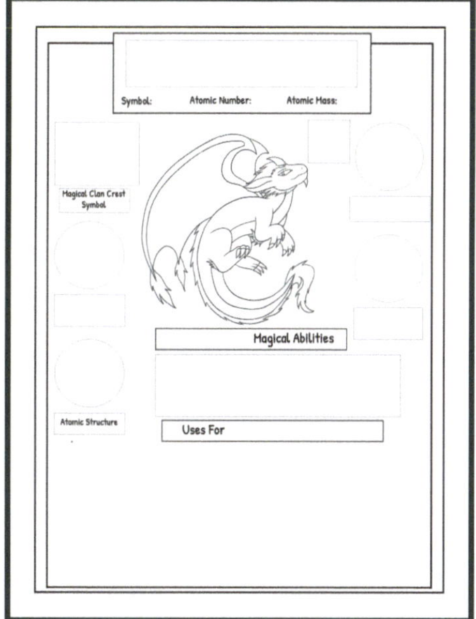

Blank Dragon Element Card

Magical Dragon Elemental Research Sheet

Before starting your Magical Dragon Elemental graphics page, do some research on your chosen element.

Name of Magical Dragon:

Dragon's Magic Power Based on the Element's Properties:

Magical Clan Crest Symbol:

Element Name:

Element Symbol:

Atomic Number:

Atomic Mass:

What year and where was this Element discovered?

Who discovered this Element?

Element Group:

Element Period:

Element Family Name:

State of Element At Room Temperature:

What is Element Mined or Extracted From?

Is Element Magnetic?

Does Element Conduct Electricity?

Where is the Element commonly found in Nature?

What is 1 alloy of the Element? How used?

What is 1 compound of the Element? How used?

Name the most common use for this Element:

Name a little known use for this Element:

Name one more use for this Element:

Interesting and Fun Facts:

Blank Research Sheet

Using the sample Magical Elemental cards provided, have students select an element from the Periodic Table and a Magical Elemental Card Blank to create their own Magical Elemental Card. The blank and sample cards do not have to match.

You will receive a pdf containing either 26 unicorn or 26 dragon sample cards and blanks to be printed on 8 1/2 x 11 sized paper or card stock. The pdf also contains a Magical Elemental Research Sheet for the students to work on before creating their unique Periodic Table Elemental. They will also write a short paragraph describing their Unicorn or Dragon Elemental from that research.

Get These Fun Elemental Periodic Table Activity Sheets at *MagicalPTElements.com*

Don't Forget To Get A Tee Shirt

Featuring Your Favorite of the

118 Elements from the Periodic Table

Available in adult and kid sizes in many colors.

https://amzn.to/47NVZWN

This is the Bohrium-Blaadlak Tee Shirt Graphic

Get it at https://www.amazon.com/dp/B0DND92LD5

Dear Reader

I hope the "No Metal No Magic Element 2 — Hetha, from The Magical Elements of the Periodic Table Series Presents Helium" with illustrations by Pranavva et al, has helped you learn some fun and interesting things about the magic of the element, Helium.

This is one of what will eventually be 118 books featuring periodic table elements presented by unicorns, dragons, wizards, knights and goblins. Keep checking regularly. Every one of the elements are amazing and very necessary to our

Techno-magical.

A lot of research went into every page of this book as well as the Magical Elements of the Periodic Table Books. There are just too many references to publish in this book but you can read and research them all at MagicalPTElements.com/MAUPT or /MDAPT or /MW1PT or / MW2PT or /MAKPT. There, you can also access book related activity sheets and games to help make the learning process more fun.

Get ready made trading cards, lapel pins, tee shirts and more based on this book from Sybrina Publishing's No Metal No Magic Collection at Zazzle - **http://bit.ly/3km64Wg**

Would you like a 24" x 36" poster of the Elemental-Themed Periodic Table in this book? The best place to get it is at **https://bit.ly/49QMxBT** They have the sharpest images of any other poster printer around.

The Magical Elements of the Periodic Table books came into existence because of my Blue Unicorn—Journey To Osm books. If it weren't for their magical powers, based on the properties of the metals of their horns and hooves, I would have never come up with the idea to relate magical creatures to the periodic table. There's a metal horn unicorn story for every age group and they are all available at MagicalPTElements.com

If you enjoyed this book
please leave a nice review
at your favorite online book site.

No Metal No Magic

Song Lyrics

No metal, no Magic

No metal, no Magic

I can think of nothing more tragic

Than to have no metal or no magic

Metal makes everything magical.

Just ask a unicorn. . .

Preferably, one with a metal horn.

They'd say No metal, No magic.

Metal makes everything techno magical.

No metal, No magic

for two-leggers or unicorns.

No metal, No magic

Metal makes everything techno magical.

No metal, No magic

It's techno magical.

No metal, No magic

It might be very hard to believe but with

No metal, No magic

There'd be no technology.

No metal, No magic

Blue Unicorn

Ebooks, Audio and Print Books Available at all online book stores.

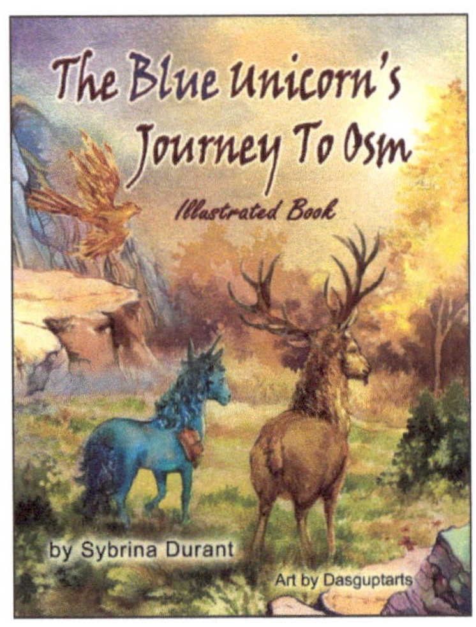

Illustrated Book

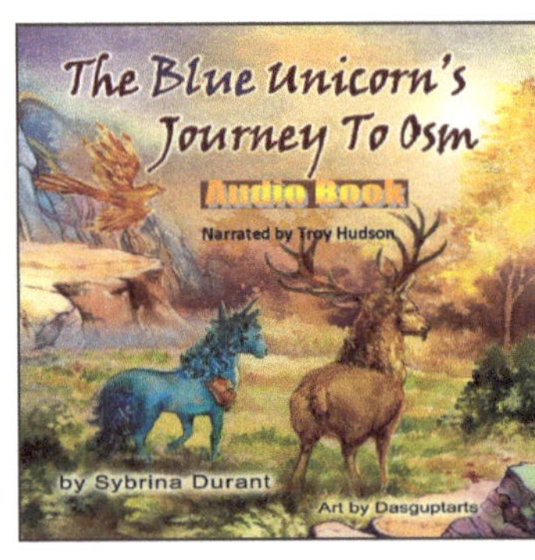

Audio Book

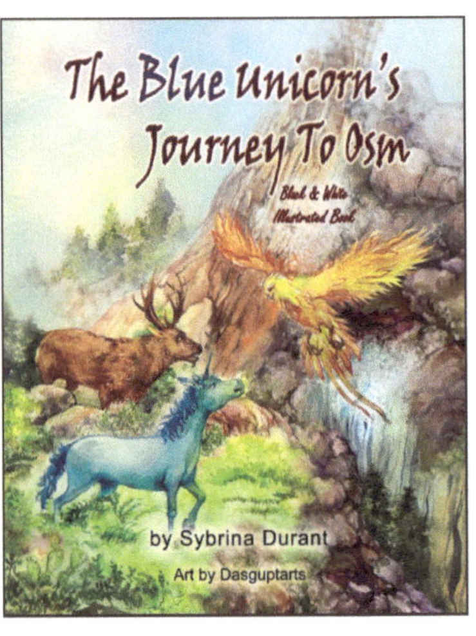

'Read & Color' Book For Teens

Unicorn Periodic Table Book

Fantasy Novel

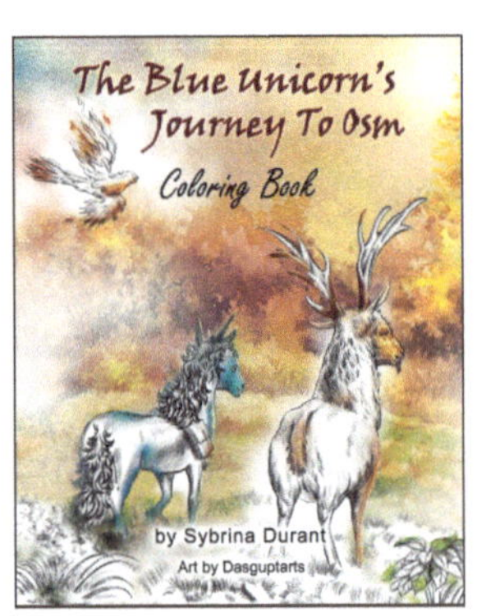

Coloring Book & Character Introduction

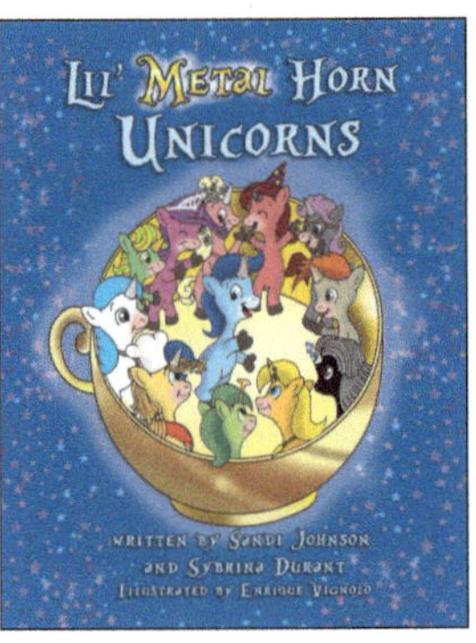

Picture Book For Kids

Get these and more at

MagicalPTElements.com

and Sybrina.com

Journey To Osm—The Blue Unicorn's Tale

Back then, most places throughout MarBryn had wizards and sorcerers of some ability, or another. Most were trained in the ways of magical arts by the unicorns as part of their outreach program. Some two-leggers developed practical magical skills like making delicious feasts of tasty food appear out of thin air or purifying murky water around the land.

Others went through more extensive training to learn battle magic—like shooting powerful streams of energy from their swords.

Some were taught the art of holding the glow of the sun in magical globes, bringing light into the dark of night. These magical lights warded off the evil beings that were new and frightening products of dark magic.

With the rise of the sorcerer Magh, magical defense arts had become more important. The highest level of magical training involved sensing when others were in danger and learning to see into the future. Very few two-leggers ever reached that level because magic wasn't inherent in them the way it was for the unicorns. The metal of their horns and hooves were part of them as well as the very makeup of their blood, but two-leggers relied on learned magic via potions, charms, and incantations that required help from ingredients and forces more mystical in nature than any two-legger was ever born to be. Of course, controlling magic and projecting your intentions went far beyond merely following a recipe of sorts. It took being in touch with nature and the various elements to get the response a wizard desired. Much trial and error went into it, as well as faith and trust and the motives of the spell caster.

Magic was and is a practice that is never quite perfected even for the unicorns who must continue to hone and learn how to harness their powers. On rare occasion a wizard and unicorn had formed enough trust and a steadfast bond that prompted the unicorn to gift the wizard with a wand or staff embedded with the smallest sliver of metal from one of their hooves. This was rare but had happened and of course so had the desire for more power. Magh wasn't the first sorcerer with lust for more and throughout history there had been a handful of heinous acts against unicorns from those seeking their magic. Prior to Magh those wizards had failed to circumvent the protections nature had infused unicorn magic with so even after harvesting metal from their horns or hooves these sorcerers had gone mad trying to bend the will of nature and actually use their ill-gotten gains.

Magh, too, had gone mad or perhaps he'd already been so, but somehow he'd managed to harness the metal he harvested from his victims and continued to grow stronger rather than completely lose his mind like the others. How exactly remained a mystery to the unicorns and everyone else.

The wizards who'd honed their craft with the help and blessing of the unicorns tried to solve the riddle and even the best of them failed to uncover that secret. Some of MarBryn's natives took to magic naturally, while others struggled with the concept but no matter how adept they were. All of them fought valiantly against Magh's magic because with even a shred of knowledge they understood how the power shift would ultimately play out. They fought to the end but, in the end, only one sorcerer remained in MarBryn and now, Magh was in total control. But still, he was not satisfied. He wanted to control every living creature in the land.